Selective Detectors

CHEMICAL ANALYSIS

A SERIES OF MONOGRAPHS ON
ANALYTICAL CHEMISTRY AND ITS APPLICATIONS

Editor
J. D. WINEFORDNER

VOLUME 131

A WILEY-INTERSCIENCE PUBLICATION

JOHN WILEY & SONS, INC.

New York / Chichester / Brisbane / Toronto / Singapore

Selective Detectors

Environmental, Industrial, and Biomedical Applications

Edited by
ROBERT E. SIEVERS

University of Colorado at Boulder
Boulder, Colorado

A WILEY-INTERSCIENCE PUBLICATION

JOHN WILEY & SONS, INC.

New York / Chichester / Brisbane / Toronto / Singapore

This text is printed on acid-free paper.

Library of Congress Cataloging in Publication Data:

Sievers, Robert E.
    Selective detectors : environmental, industrial, and biomedical
applications / Robert E. Sievers.
        p.   cm. — (Chemical analysis)
    "A Wiley-Interscience publication."
    Includes index.
    ISBN 0-471-01343-9
    1. Chemical detectors.   I. Title.   II. Series.
TP159.C46S54   1995
660—dc20   94-44366

Printed in the United States of America

10 9 8 7 6 5 4 3 2 1

CONTENTS

CONTRIBUTORS   xiii

PREFACE   xv

CUMULATIVE LISTING OF VOLUMES IN SERIES   xvii

CHAPTER 1   **SULFUR-SELECTIVE CHEMILUMINESCENCE DETECTION IN CHROMATOGRAPHY**   1

Thomas B. Ryerson and Robert E. Sievers

1.1. Introduction   1

     1.1.1. Use of Sulfur Chemiluminescence in Chromatography   3

1.2. Flame Photometry   3

     1.2.1. Chemiluminescence Mechanism   3

     1.2.2. Chemiluminescence Efficiency   5

     1.2.3. Detector Design   5

         1.2.3.1. Single-Flame FPD   5

         1.2.3.2. Dual-Flame FPD   5

     1.2.4. Response Characteristics and Applications   7

1.3. Fluorine-Induced Chemiluminescence   8

     1.3.1. Chemiluminescence Mechanism   8

     1.3.2. Chemiluminescence Efficiency   9

     1.3.3. Detector Design   10

     1.3.4. Response Characteristics and Applications   10

1.4. Ozone-Induced Chemiluminescence   12

     1.4.1. Introduction   12

     1.4.2. Direct Reaction with Ozone   13

         1.4.2.1. Chemiluminescence Mechanism   13

         1.4.2.2. Chemiluminescence Efficiency   14

         1.4.2.3. Detector Design   14

         1.4.2.4. Response Characteristics and Applications   14

1.4.3. Flame Conversion   15

    1.4.3.1. Chemiluminescence Mechanism   15

    1.4.3.2. Chemiluminescence Efficiency   16

    1.4.3.3. Detector Design   16

    1.4.3.4. Response Characteristics and
           Applications   18

    1.4.3.5. Flameless Combustion   19

1.4.4. Catalytic Thermionic Conversion   19

    1.4.4.1. Method of SO Formation   19

    1.4.4.2. Chemiluminescence Efficiency   19

    1.4.4.3. Detector Design   19

    1.4.4.4. Response Characteristics and
           Applications   20

1.4.5. Aqueous Phase Oxidation   22

    1.4.5.1. Mechanism of SO Formation   22

    1.4.5.2. Chemiluminescence Efficiency   24

    1.4.5.3. Detector Design   24

    1.4.5.4. Response Characteristics and
           Applications   24

References   30

**CHAPTER 2   FLAMELESS SULFUR CHEMILUMINESCENCE
DETECTION**   35
Randall Lee Shearer

2.1. Introduction   35

  2.1.1. Need for Sulfur Compound Analysis in the
       Petroleum Industry   36

  2.1.2. Sulfur-Selective Detectors   37

2.2. Principle of Operation of the Sulfur
     Chemiluminescence Detector   38

  2.2.1. Combustion Chemistry of Sulfur   38

  2.2.2. Chemiluminescence Detection   40

2.3. Flameless Sulfur Chemiluminescence Detection   42

  2.3.1. Differences from Flame-Based Sulfur
       Chemiluminescence Detection   43

  2.3.2. Operation and Performance   45

    2.3.2.1. Sensitivity   48

    2.3.2.2. Selectivity   51

    2.3.2.3. Relative Response Factors   52

2.3.2.4.	*Linearity*	54
2.3.2.5.	*Precision*	54
2.3.2.6.	*Other Factors*	57

2.4. Applications   59
  2.4.1. Gas Chromatography   59
  2.4.2. Supercritical Fluid Chromatography   64
  2.4.3. Liquid Chromatography   66
2.5. Concluding Remarks   66
References   67

**CHAPTER 3   SELECTIVE DETECTION TECHNIQUES BASED
ON NITRIC OXIDE/OZONE
CHEMILUMINESCENCE   71**
Andrew J. Dunham and Robert E. Sievers

3.1. Introduction   71
  3.1.1. History   71
  3.1.2. Principle of Operation   72
  3.1.3. Summary of Instrument Development   74
3.2. Analyses of Aqueous NO, Nitrite Ion, and Nitrate
     Ion by NO Chemiluminescence   75
  3.2.1. Introductory Remarks   75
  3.2.2. Analysis of Nitrite and Nitrate Ions in
         Aqueous Samples   75
  3.2.3. Analysis of NO and Nitrite Ion in
         Seawater   76
  3.2.4. Analysis of Nitrate Ion in Atmospheric
         Particles   77
  3.2.5. Flow Injection Analysis Techniques for
         Nitrite Ion in Aqueous Samples   77
3.3. Analysis of Nitrogen-Containing Organic
     Compounds in Liquid Samples by NO
     Chemiluminescence   81
  3.3.1. Background   81
  3.3.2. Analysis of *N*-Nitroso Compounds   81
    3.3.2.1. *Principle of Operation*   81
       3.3.2.1.1. *Thermal Energy Analysis*   81
       3.3.2.1.2. *Reductive Cleavage of
                   Nitrosamines*   83
    3.3.2.2. *Applications*   83

3.3.3. Analysis of Organonitrogen Compounds   84

    3.3.3.1. Principle of Operation   84

    3.3.3.2. Applications   84

3.4. Analysis of Atmospheric Nitrogen-Containing
Compounds   85

3.4.1. Analysis of NO in the Atmosphere   85

3.4.2. Thermal Conversion of NO_x and NH_3 for
Detection   86

3.4.3. Nonthermal Conversion of NO_2 to NO for
Detection   88

    3.4.3.1. Reduction of NO_2 to NO over $FeSO_4$ 88

    3.4.3.2. Photolysis of NO_2 to NO   88

3.4.4. Catalytic Conversion of NO_y to NO for
Detection   89

3.5. The Redox Chemiluminescence Detector   91

3.5.1. Background   91

3.5.2. Principle of Operation   91

3.5.3. Applications   92

    3.5.3.1. Gas Chromatography   92

    *3.5.3.2. Supercritical Fluid Chromatography
and Liquid Chromatography*   93

References   95

CHAPTER 4   THE OXYGEN-SENSITIVE DETECTOR O-FID   99
Ulrich K. Gökeler

4.1. Introduction   99

4.2. Principles of Operation   101

4.3. Design   103

4.3.1. The Cracking Reactor   103

4.3.2. The Methanizer   105

4.3.3. Gas Supply   106

4.4. Performance   107

4.4.1. Analytical Parameters   107

4.4.2. Dynamic Range   107

4.4.3. Linearity   108

4.4.4. Sensitivity   108

4.4.5. Selectivity   109

4.4.6. Response Factors   109

4.4.7. Reproducibility   110

4.4.8. The Cracking Temperature   111

4.4.9. The Cracking Reactor and Methanizer
      Poisoning   113

4.5. Applications   113

    4.5.1. Automotive Applications   114

    4.5.2. Natural Products   115

4.6. Some Remarks as to O-FID Operation   122

4.7. Summary   124

References   124

**CHAPTER 5   METAL SELECTIVITY BY ION MOBILITY
DISCRIMINATION IN A FLAME IONIZATION
DETECTOR**   **127**

Michael M. Gallagher and Herbert H. Hill, Jr.

5.1. Overview of Detector Response   127

5.2. Experimental Considerations   129

    5.2.1. Instrumentation   129

    5.2.2. Reagents   130

    5.2.3. Biasing Electrode Experiments   131

    5.2.4. Computer Modeling Experiments   132

5.3. Results and Discussion   132

    5.3.1. Collar-Biasing Experiments   132

    5.3.2. Wall-Biasing Experiments   134

    5.3.3. Computer Modeling Experiments   139

5.4. Conclusion   141

References   142

**CHAPTER 6   ELEMENT-SELECTIVE GAS
CHROMATOGRAPHIC DETECTION BY
ATOMIC PLASMA EMISSION SPECTROSCOPY   143**

Peter C. Uden

6.1. Introduction   143

    6.1.1. The Selective Chromatographic Detector   143

    6.1.2. The Element-Selective Detector   144

6.2. Overview of Instrumentation   144

    6.2.1. Types of Atomic Plasma Emission
        Chromatographic Detector   144

6.2.1.1.   The Microwave-Induced Electrical
           Discharge Plasma (MIP) Detector   145

6.2.1.2.   Other Low-Power Plasma
           Chromatographic Detectors   146

6.2.1.3.   Higher Power Argon Plasma Detectors   147

6.2.2. Analytical Information from
        Chromatography–Atomic Emission
        Detection (C/AED)   148

6.2.3. Plasma Interfacing with Capillary GC   150

6.3. Gas Chromatography–Atomic Emission Detection   150

6.3.1. GC/AED of Nonmetallic Elements   150

6.3.1.1.   Reduced Pressure Plasmas   151

6.3.1.2.   Atmospheric Pressure Plasmas   153

6.3.2. GC/AED of Metals   160

6.3.2.1.   GC/AED of Main Group Metallic
           Compounds   160

6.3.2.2.   GC/AED of Transition Metal
           Compounds   163

6.4. The Future Development of Atomic Emission
      Chromatographic Detection   167

References   167

CHAPTER 7   SELECTIVE CHROMATOGRAPHIC
            DETECTION BY PLASMA MASS
            SPECTROMETRY   171
            Francine A. Byrdy and Joseph A. Caruso

7.1. Introduction   171

7.1.1. Speciation   172

7.1.1.1.   Arsenic   172

7.1.1.2.   Cadmium   172

7.1.1.3.   Chromium   172

7.1.1.4.   Gold   173

7.1.1.5.   Lead   173

7.1.1.6.   Iodine   173

7.1.1.7.   Iron   173

7.1.1.8.   Mercury   174

7.1.1.9.   Selenium   174

7.1.1.10. Tellurium   174

       7.1.1.11. Tin   174
   7.1.2. Plasma MS   175
7.2. High-Performance Liquid Chromatography   176
   7.2.1. Reversed-Phase Liquid Chromatography   177
       7.2.1.1. Applications   178
   7.2.2. Ion-Pairing Chromatography   179
       7.2.2.1. Applications   180
   7.2.3. Micellar Liquid Chromatography   182
       7.2.3.1. Applications   183
   7.2.4. Ion-Exchange Chromatography   184
       7.2.4.1. Applications   184
   7.2.5. Size-Exclusion Chromatography   186
       7.2.5.1. Applications   186
   7.2.6. Comparison Studies   186
       7.2.6.1. Special Studies   188
           7.2.6.1.1. On-Line Preconcentration of Trace Metals   188
           7.2.6.1.2. Gradient Elution   188
           7.2.6.1.3. Multielement Time-Resolved Analysis   188
           7.2.6.1.4. Direct Injection Nebulization   189
           7.2.6.1.5. Argon Chloride Interference Elimination   189
   7.2.7. Future Developments   190
7.3. Gas Chromatography   190
   7.3.1. GC/MIP/MS   191
       7.3.1.1. Low-Pressure GC/MIP/MS   192
           7.3.1.1.1. Applications   192
       7.3.1.2. Atmospheric Pressure GC/MIP/MS   195
           7.3.1.2.1. Applications   195
   7.3.2. GC/ICP/MS   196
   7.3.3. Future Directions   197
7.4. Supercritical Fluid Chromatography   198
   7.4.1. Applications   200
   7.4.2. Future Developments   202
7.5. Conclusions   203

7.6. Summary   203
References   204

**CHAPTER 8   PEROXYOXALATE CHEMILUMINESCENCE:
MECHANISM AND ANALYTICAL DETECTION   209**
Andrew G. Hadd and John W. Birks

8.1. Reaction Mechanism   210
    8.1.1. Kinetics Studies Indicating Multiple
    Reaction Pathways   213
    8.1.2. Mechanistic Information Obtained from
    Measurements of Background Emission   216
    8.1.3. Mechanistic Information Obtained from
    Photochemical Initiation   217
8.2. Optimization of Peroxyoxalate Chemiluminescence
for Analytical Applications   221
    8.2.1. Optimization of Reaction Conditions   222
    8.2.2. Optimization of Delivery and Mixing of the
    PO–CL Reagents   224
    8.2.3. Optimization of Light Detection   226
8.3. Analytical Applications   227
    8.3.1. Detection of Fluorescent Compounds   228
    8.3.2. Detection of Compounds Derivatized with
    a Fluorescent Tag   229
    8.3.3. Detection of Hydrogen Peroxide Produced
    in a Postcolumn Reaction   230
    8.3.4. Detection of Oxalates   235
    8.3.5. Detection Based on Enhancement of the
    Background Signal   235
    8.3.6. Analysis by Quenched Peroxyoxalate
    Chemiluminescence   235
8.4. Conclusions and Future Directions   236
References   238

**CHAPTER 9   TALES OF A RELUCTANT INSTRUMENT
MAKER   241**
James E. Lovelock

INDEX   251

CONTRIBUTORS

John W. Birks, Department of Chemistry and Biochemistry, Cooperative Institute for Research in Environmental Sciences (CIRES), University of Colorado at Boulder, Boulder, Colorado

Francine A. Byrdy, Department of Chemistry, University of Cincinnati, Cincinnati, Ohio

Joseph A. Caruso, Department of Chemistry, University of Cincinnati, Cincinnati, Ohio

Andrew J. Dunham, Department of Chemistry and Biochemistry, Cooperative Institute for Research in Environmental Sciences (CIRES), University of Colorado at Boulder, Boulder, Colorado

Michael M. Gallagher, Department of Chemistry, Washington State University, Pullman, Washington

Ulrich K. Gökeler, ES Industries, Division for Siemens Chromatographs, Berlin, New Jersey

Andrew G. Hadd, Department of Chemistry, Cooperative Institute of Environmental Sciences (CIRES), University of Colorado at Boulder, Boulder, Colorado

Herbert H. Hill, Jr., Department of Chemistry, Washington State University, Pullman, Washington

James E. Lovelock, Coombe Mill, St. Giles on the Heath, Launceton, Cornwall, England

Thomas B. Ryerson, Department of Chemistry and Biochemistry, Cooperative Institute for Research in Environmental Sciences (CIRES), University of Colorado at Boulder, Boulder, Colorado

Randall Lee Shearer, Sievers Instruments, Boulder, Colorado

Robert E. Sievers, Global Change and Environmental Quality Program, University of Colorado at Boulder, Boulder, Colorado

Peter C. Uden, Department of Chemistry, University of Massachusetts, Amherst, Massachusetts

PREFACE

The development of selective detectors has had a profound effect on analytical, environmental, and biomedical sciences. The detection and measurement of trace levels of certain compounds in the presence of much larger concentrations of other compounds has become possible because of the distinctiveness of the signals generated. When the most sensitive and selective sensor is coupled with the best in separation methodology, the result may be the discovery of previously unrecognized phenomena. In the 1960s, the development of selective mass spectrometric and electron capture detectors, coupled with gas chromatography, led to the realization that certain halogenated compounds were so stable that they were being transported long distances and accumulating in the biosphere and atmosphere. In the 1980s and 1990s, the discovery of the many roles of nitric oxide and its products and precursors in physiological systems, as well as in biogeochemical processes, has been greatly facilitated by the development of highly selective and sensitive chemiluminescence methods.

This book focuses on the more recent advances in the development of selective detectors and concludes with a fascinating historical account by Dr. James Lovelock of some of his early work on selective measurement devices. This book is dedicated to Dr. Lovelock and the other pioneers in this field, on whose shoulders we stand. When I began my research in the 1950s, I marveled at the willingness of senior scientists to provide advice and encouragement to fledgling researchers. By reading the Lovelock chapter, one can recapture today the sense of how those pioneers have inspired and supported us. Dr. Lovelock invented the electron capture detector, which he and so many of us have used to learn about the Earth and its living systems. Jim Lovelock and others have suggested that Earth processes can be looked at as parts of a self-regulating system, as advanced in the Gaia hypothesis.

In my artistic life, I carve marble statues, and I have also dedicated a 1,000-pound monument entitled "Gaia: Earth Mother," to Dr. Lovelock. It has been exhibited in various places, most recently in Boulder Central Park, near the lecture hall where Dr. Lovelock delivered a particularly inspiring lecture on "geophysiology" and Gaia.

I am grateful to many people who have made the preparation of this book possible. Ms. Rosella Chávez deserves my deepest thanks for her hard work and superb support. Three of my senior graduate students, Dr. Thomas Ryerson, Dr. Andrew Dunham, and Ms. Kris Hansen, assisted with the editing. The approval of some of the organizers of Pittcon, Dr. Vince Conrad, Dr. Johannes Coetzee, and Dr. W. R. Howe, to publish the 1994 Pittsburgh Conference lecture by Lovelock is greatly appreciated. Finally, the understanding and support of my wife, Nancy Klindera, over the past 33 years has been immeasurable and greatly valued.

ROBERT E. SIEVERS

Boulder, Colorado
January 1995

CHEMICAL ANALYSIS

A SERIES OF MONOGRAPHS ON
ANALYTICAL CHEMISTRY AND ITS APPLICATIONS

J. D. Winefordner, *Series Editor*

Vol.   1. **The Analytical Chemistry of Industrial Poisons, Hazards, and Solvents.** *Second Edition.* By the late Morris B. Jacobs

Vol.   2. **Chromatographic Adsorption Analysis.** By Harold H. Strain (*out of print*)

Vol.   3. **Photometric Determination of Traces of Metals.** *Fourth Edition*
Part I: General Aspects. By E. B. Sandell and Hiroshi Onishi
Part IIA: Individual Metals, Aluminum to Lithium. By Hiroshi Onishi
Part IIB: Individual Metals, Magnesium to Zirconium. By Hiroshi Onishi

Vol.   4. **Organic Reagents Used in Gravimetric and Volumetric Analysis.** By John F. Flagg (*out of print*)

Vol.   5. **Aquametry: A Treatise on Methods for the Determination of Water.** *Second Edition* (*in three parts*). By John Mitchell, Jr. and Donald Milton Smith

Vol.   6. **Analysis of Insecticides and Acaricides.** By Francis A. Gunther and Roger C. Blinn (*out of print*)

Vol.   7. **Chemical Analysis of Industrial Solvents.** By the late Morris B. Jacobs and Leopold Schetlan

Vol.   8. **Colorimetric Determination of Nonmetals.** *Second Edition.* Edited by the late David F. Boltz and James A. Howell

Vol.   9. **Analytical Chemistry of Titanium Metals and Compounds.** By Maurice Codell

Vol. 10. **The Chemical Analysis of Air Pollutants.** By the late Morris B. Jacobs

Vol. 11. **X-Ray Spectrochemical Analysis.** *Second Edition.* By L. S. Birks

Vol. 12. **Systematic Analysis of Surface-Active Agents.** *Second Edition.* By Milton J. Rosen and Henry A. Goldsmith

Vol. 13. **Alternating Current Polarography and Tensammetry.** By B. Breyer and H. H. Bauer

Vol. 14. **Flame Photometry.** By R. Herrmann and J Alkemade

Vol. 15. **The Titration of Organic Compounds** (*in two parts*). By M. R. F. Ashworth

Vol. 16. **Complexation in Analytical Chemistry: A Guide for the Critical Selection of Analytical Methods Based on Complexation Reactions.** By the late Anders Ringbom

Vol. 17. **Electron Probe Microanalysis.** *Second Edition.* By L. S. Birks

Vol. 18. **Organic Complexing Reagents: Structure, Behavior, and Application to Inorganic Analysis.** By D. D. Perrin

Vol.   19.   **Thermal Analysis.** *Third Edition.* By Wesley Wm. Wendlandt

Vol.   20.   **Amperometric Titrations.** By John T. Stock

Vol.   21.   **Reflectance Spectroscopy.** By Wesley Wm. Wendlandt and Harry G. Hecht

Vol.   22.   **The Analytical Toxicology of Industrial Inorganic Poisons.** By the late Morris B. Jacobs

Vol.   23.   **The Formation and Properties of Precipitates.** By Alan G. Walton

Vol.   24.   **Kinetics in Analytical Chemistry.** By Harry B. Mark, Jr. and Garry A. Rechnitz

Vol.   25.   **Atomic Absorption Spectroscopy.** *Second Edition.* By Morris Slavin

Vol.   26.   **Characterization of Organometallic Compounds** (*in two parts*). Edited by Minoru Tsutsui

Vol.   27.   **Rock and Mineral Analysis.** *Second Edition.* By Wesley M. Johnson and John A. Maxwell

Vol.   28.   **The Analytical Chemistry of Nitrogen and Its Compounds** (*in two parts*). Edited by C. A. Streuli and Philip R. Averell

Vol.   29.   **The Analytical Chemistry of Sulfur and Its Compounds** (*in three parts*). By J. H. Karchmer

Vol.   30.   **Ultramicro Elemental Analysis.** By Günther Tölg

Vol.   31.   **Photometric Organic Analysis** (*in two parts*). By Eugene Sawicki

Vol.   32.   **Determination uf Organic Compounds: Methods and Procedures.** By Frederick T. Weiss

Vol.   33.   **Masking and Demasking of Chemical Reactions.** By D. D. Perrin

Vol.   34.   **Neutron Activation Analysis.** By D. De Soete, R. Gijbels, and J. Hoste

Vol.   35.   **Laser Raman Spectroscopy.** By Marvin C. Tobin

Vol.   36.   **Emission Spectrochemical Analysis.** By Morris Slavin

Vol.   37.   **Analytical Chemistry of Phosphorus Compounds.** Edited by M. Halmann

Vol.   38.   **Luminescence Spectrometry in Analytical Chemistry.** By J. D. Winefordner, S. G. Schulman and T. C. O'Haver

Vol.   39.   **Activation Analysis with Neutron Generators.** By Sam S. Nargolwalla and Edwin P. Przybylowicz

Vol.   40.   **Determination of Gaseous Elements in Metals.** Edited by Lynn L. Lewis, Laben M. Melnick, and Ben D. Holt

Vol.   41.   **Analysis of Silicones.** Edited by A. Lee Smith

Vol.   42.   **Foundations of Ultracentrifugal Analysis.** By H. Fujita

Vol.   43.   **Chemical Infrared Fourier Transform Spectroscopy.** By Peter R. Griffiths

Vol.   44.   **Microscale Manipulations in Chemistry.** By T. S. Ma and V. Horak

Vol.   45.   **Thermometric Titrations.** By J. Barthel

Vol.   46.   **Trace Analysis: Spectroscopic Methods for Elements.** Edited by J. D. Winefordner

Vol.   47.   **Contamination Control in Trace Element Analysis.** By Morris Zief and James W. Mitchell

Vol.   48.   **Analytical Applications of NMR.** By D. E. Leyden and R. H. Cox

Vol.   49.   **Measurement of Dissolved Oxygen.** By Michael L. Hitchman

Vol.   50.   **Analytical Laser Spectroscopy.** Edited by Nicolo Omenetto

Vol.   51.   **Trace Element Analysis of Geological Materials.** By Roger D. Reeves and Robert R. Brooks

Vol. 52. **Chemical Analysis by Microwave Rotational Spectrscopy.** By Ravi Varma and Lawrence W. Hrubesh

Vol. 53. **Information Theory As Applied to Chemical Analysis.** By Karel Eckschlager and Vladimir Štěpánek

Vol. 54. **Applied Infrared Spectroscopy: Fundamentals, Techniques, and Analytical Problem-solving.** By A. Lee Smith

Vol. 55. **Archaeological Chemistry.** By Zvi Goffer

Vol. 56. **Immobilized Enzymes in Analytical and Clinical Chemistry.** By P. W. Carr and L. D. Bowers

Vol. 57. **Photoacoustics and Photoacoustic Spectroscopy.** By Allan Rosencwaig

Vol. 58. **Analysis of Pesticide Residues.** Edited by H. Anson Moye

Vol. 59. **Affinity Chromatography.** By William H. Scouten

Vol. 60. **Quality Control in Analytical Chemistry.** *Second Edition.* By G. Kateman and L. Buydens

Vol. 61. **Direct Characterization of Fineparticles.** By Brian H. Kaye

Vol. 62. **Flow Injection Analysis.** By J. Ruzicka and E. H. Hansen

Vol. 63. **Applied Electron Spectroscopy for Chemical Analysis.** Edited by Hassan Windawi and Floyd Ho

Vol. 64. **Analytical Aspects of Environmental Chemistry.** Edited by David F. S. Natusch and Philip K. Hopke

Vol. 65. **The Interpretation of Analytical Chemical Data by the Use of Cluster Analysis.** By D. Luc Massart and Leonard Kaufman

Vol. 66. **Solid Phase Biochemistry: Analytical and Synthetic Aspects.** Edited by William H. Scouten

Vol. 67. **An Introduction to Photoelectron Spectroscopy.** By Pradip K. Ghosh

Vol. 68. **Room Temperature Phosphorimetry for Chemical Analysis.** By Tuan Vo-Dinh

Vol. 69. **Potentiometry and Potentiometric Titrations.** By E. P. Serjeant

Vol. 70. **Design and Application of Process Analyzer Systems.** By Paul E. Mix

Vol. 71. **Analysis of Organic and Biological Surfaces.** Edited by Patrick Echlin

Vol. 72. **Small Bore Liquid Chromatography Columns: Their Properties and Uses.** Edited by Raymond P. W. Scott

Vol. 73. **Modern Methods of Particle Size Analysis.** Edited by Howard G. Barth

Vol. 74. **Auger Electron Spectroscopy.** By Michael Thompson, M. D. Baker, Alec Christie, and J. F. Tyson

Vol. 75. **Spot Test Analysis: Clinical, Environmental, Forensic and Geochemical Applications.** By Ervin Jungreis

Vol. 76. **Receptor Modeling in Environmental Chemistry.** By Philip K. Hopke

Vol. 77. **Molecular Luminescence Spectroscopy: Methods and Applications** (*in three parts*). Edited by Stephen G. Schulman

Vol. 78. **Inorganic Chromatographic Analysis.** Edited by John C. MacDonald

Vol. 79. **Analytical Solution Calorimetry.** Edited by J. K. Grime

Vol. 80. **Selected Methods of Trace Metal Analysis: Biological and Environmental Samples.** By Jon C. VanLoon

Vol. 81. **The Analysis of Extraterrestrial Materials.** By Isidore Adler

Vol. 82. **Chemometrics.** By Muhammad A. Sharaf, Deborah L. Illman, and Bruce R. Kowalski

Vol.   83.   **Fourier Transform Infrared Spectrometry.** By Peter R. Griffiths and James A. de Haseth

Vol.   84.   **Trace Analysis: Spectroscopic Methods for Molecules.** Edited by Gary Christian and James B. Callis

Vol.   85.   **Ultratrace Analysis of Pharmaceuticals and Other Compounds of Interest.** Edited by S. Ahuja

Vol.   86.   **Secondary Ion Mass Spectrometry: Basic Concepts, Instrumental Aspects, Applications and Trends.** By A. Benninghoven, F. G. Rüdenauer, and H. W. Werner

Vol.   87.   **Analytical Applications of Lasers.** Edited by Edward H. Piepmeier

Vol.   88.   **Applied Geochemical Analysis.** By C. O. Ingamells and F. F. Pitard

Vol.   89.   **Detectors for Liquid Chromatography.** Edited by Edward S. Yeung

Vol.   90.   **Inductively Coupled Plasma Emission Spectroscopy: Part I: Methodology, Instrumentation, and Performance; Part II: Applications and Fundamentals.** Edited by J. M. Boumans

Vol.   91.   **Applications of New Mass Spectrometry Techniques in Pesticide Chemistry.** Edited by Joseph Rosen

Vol.   92.   **X-Ray Absorption: Principles, Applications, Techniques of EXAFS, SEXAFS, and XANES.** Edited by D. C. Konnigsberger

Vol.   93.   **Quantitative Structure–Chromatographic Retention Relationships.** By Roman Kaliszan

Vol.   94.   **Laser Remote Chemical Analysis.** Edited by Raymond M. Measures

Vol.   95.   **Inorganic Mass Spectrometry.** Edited by F. Adams, R. Gijbels, and R. Van Grieken

Vol.   96.   **Kinetic Aspects of Analytical Chemistry.** By Horacio A. Mottola

Vol.   97.   **Two-Dimensional NMR Spectroscopy.** By Jan Schraml and Jon M. Bellama

Vol.   98.   **High Performance Liquid Chromatography.** Edited by Phyllis R. Brown and Richard A. Hartwick

Vol.   99.   **X-Ray Fluorescence Spectrometry.** By Ron Jenkins

Vol.   100.   **Analytical Aspects of Drug Testing.** Edited by Dale G. Deutsch

Vol.   101.   **Chemical Analysis of Polycyclic Aromatic Compounds.** Edited by Tuan Vo-Dinh

Vol.   102.   **Quadrupole Storage Mass Spectrometry.** By Raymond E. March and Richard J. Hughes

Vol.   103.   **Determination of Molecular Weight.** Edited by Anthony R. Cooper

Vol.   104.   **Selectivity and Detectability Optimizations in HPLC.** By Satinder Ahuja

Vol.   105.   **Laser Microanalysis.** By Lieselotte Moenke-Blankenburg

Vol.   106.   **Clinical Chemistry.** Edited by E. Howard Taylor

Vol.   107.   **Multielement Detection Systems for Spectrochemical Analysis.** By Kenneth W. Busch and Marianna A. Busch

Vol.   108.   **Planar Chromatography in the Life Sciences.** Edited by Joseph C. Touchstone

Vol.   109.   **Fluorometric Analysis in Biomedical Chemistry: Trends and Techniques Including HPLC Applications.** By Norio Ichinose, George Schwedt, Frank Michael Schnepel, and Kyoko Adochi

Vol.   110.   **An Introduction to Laboratory Automation.** By Victor Cerdá and Guillermo Ramis

Vol.   111.   **Gas Chromatography: Biochemical, Biomedical, and Clinical Applications.** Edited by Ray E. Clement

Vol. 112. **The Analytical Chemistry of Silicones.** Edited by A. Lee Smith

Vol. 113. **Modern Methods of Polymer Characterization.** Edited by Howard G. Barth and Jimmy W. Mays

Vol. 114. **Analytical Raman Spectroscopy.** Edited by Jeannette Graselli and Bernard J. Bulkin

Vol. 115. **Trace and Ultratrace Analysis by HPLC.** By Satinder Ahuja

Vol. 116. **Radiochemistry and Nuclear Methods of Analysis.** By William D. Ehmann and Diane E. Vance

Vol. 117. **Applications of Fluorescence in Immunoassays.** By Ilkka Hemmila

Vol. 118. **Principles and Practice of Spectroscopic Calibration.** By Howard Mark

Vol. 119. **Activation Spectrometry in Chemical Analysis.** By S. J. Parry

Vol. 120. **Remote Sensing by Fourier Transform Spectrometry.** By Reinhard Beer

Vol. 121. **Detectors for Capillary Chromatography.** Edited by Herbert H. Hill and Dennis McMinn

Vol. 122. **Photochemical Vapor Deposition.** By J. G. Eden

Vol. 123. **Statistical Methods in Analytical Chemistry.** By Peter C. Meier and Richard Zund

Vol. 124. **Laser Ionization Mass Analysis.** Edited by Akos Vertes, Renaat Gijbels, and Fred Adams

Vol. 125. **Physics and Chemistry of Solid State Sensor Devices.** By Andreas Mandelis and Constantinos Christofides

Vol. 126. **Electroanalytical Stripping Methods.** By Khjena Z. Brainina and E. Neyman

Vol. 127. **Air Monitoring by Spectroscopic Techniques.** Edited by Markus W. Sigrist

Vol. 128. **Information Theory in Analytical Chemistry.** By Karel Eckschlager and Klaus Danzer

Vol. 129. **Flame Chemiluminescence Analysis by Molecular Emission Cavity Detection.** Edited by David Stiles, Anthony Calokerinos, and Alan Townshend

Vol. 130. **Hydride Generation Atomic Absorption Spectrometry.** By Jiri Dedina and Dimiter L. Tsalev

Vol. 131. **Selective Detectors: Environmental, Industrial, and Biomedical Applications.** Edited by Robert E. Sievers

CHAPTER

1

SULFUR-SELECTIVE CHEMILUMINESCENCE DETECTION IN CHROMATOGRAPHY

THOMAS B. RYERSON and ROBERT E. SIEVERS

Department of Chemistry and Biochemistry, Cooperative Institute for Research in Environmental Sciences (CIRES), and Global Change and Environmental Quality Program, University of Colorado at Boulder, Boulder, Colorado 80309

1.1. INTRODUCTION

Sulfur-containing compounds make up a large fraction of the reactive compounds that influence the chemistry of biological systems and of the atmosphere. In the atmosphere, emissions of reduced sulfur compounds from biogenic sources are being studied to better understand the chemistry leading to formation of sulfate, a major source of cloud condensation nuclei. Anthropogenic emissions of sulfur compounds to the atmosphere, largely as sulfur dioxide from fossil fuel combustion, can also lead to atmospheric sulfuric acid formation, with ensuing washout producing rainwater acidification. Sulfur compounds in biological systems are important in many contexts, from their use as pharmaceutical agents to influencing the taste and odor of foodstuffs and determining protein conformation and activities. Sulfur-containing compounds are used as fertilizers and pesticides, and as such are ubiquitous in ecosystems affected by human activity. As a result of their widespread distribution, sulfur compounds are found in a variety of complex matrices; often, the sulfur-containing analytes are present in concentrations much lower than those of the non-sulfur-containing species of that matrix. For these reasons, the development of sensitive detectors selective for sulfur-containing compounds is particularly important.

Analysis of sulfur compounds in such complex matrices presents two distinct problems: separation and identification. Separation requires the analyte be isolated from other species in the matrix that are analytically less relevant. Chromatographic separation methods are commonly employed, often

Selective Detectors, edited by Robert E. Sievers. Chemical Analysis Series, Vol. 131.
ISBN 0-471-01343-9 © 1995 John Wiley & Sons, Inc.

in combination with less selective separations such as filtration, extraction, and/or digestion. For many analyses, sample cleanup and chromatography are sufficient to separate the sulfur-containing analytes from matrix compounds. However, for more complex samples common to environmental, food, and biological analyses, the resolving power of modern chromatographic techniques may still be insufficient to fully isolate all of the sulfur-containing compounds of interest. To address the second analytical problem, identification, a detection scheme must follow the separation to confirm the presence of and quantitate analytes containing sulfur.

Because general chromatographic detectors such as the flame ionization detector (FID) response to most or all of the organic species present in a sample, producing complicated traces that may be difficult to interpret, it is often advantageous to use selective detectors in the analyses of complex mixtures. A selective detector can alleviate both problems, of separation and identification, by responding only to the analyte of interest regardless of other constituents of the matrix. A truly selective detector will require no separation of a single sulfur-containing analyte from other compounds; however, it is often convenient to couple a chromatographic step with selective detection. Thus, in some cases, chromatographic separations need not fully resolve each compound to achieve successful analysis of a sample; instead one can rely on the discrimination of the detector to distinguish and measure analytes not fully separated from other non-sulfur-containing compounds. Similarly, analyte identification is often enhanced by the use of a selective detection scheme, as the chromatographic trace is simplified to include only the compounds containing sulfur.

Sulfur-selective detection in chromatography is an active area of research and has been the topic of a recent review in this series.[1] This chapter will discuss the application of gas phase chemiluminescence (CL) to sulfur-selective detection in chromatography. Subjects covered in this context will include the chemistry involved in the generation of the various CL-active species, the emissions characteristic of each species, CL efficiency, detector design, and the application of sulfur-selective CL reactions to a chromatographic detection system. It should be noted that CL efficiency is a relative measure of intensity. In this context, a reaction in which 1 out of 10,000 excited-state analyte molecules emits a photon might be considered efficient. Nevertheless, such an "efficiency" may still permit highly sensitive detection of analytes present at only the part per billion (ppb) level.

Several nonselective CL schemes have been reported for chromatographic detection of sulfur[2-4]; because of the more general, less selective, response of these detectors, they will not be discussed in this chapter. Other chapters in this book deal in greater depth with applications of sulfur-selective detectors, and the reader is referred to these in order to develop a more complete picture of recent advances in this rapidly changing field.

1.1.1. Use of Sulfur Chemiluminescence in Chromatography

In general, three different schemes of generating chemiluminescence from sulfur-containing compounds have been employed in chromatographic detection. The first CL scheme to be described is that used in the flame photometric detector (FPD). In this case, sulfur-containing compounds undergo combustion to yield electronically excited sulfur species.[5] No added reagent gases are required. The second CL scheme is based on the reactions of molecular fluorine with reduced sulfur compounds, which produce several electronically and/or vibrationally excited species. The nature of the emitting species is dependent upon the analyte, and there may be more than one emitter for any given sulfur compound.[6-9] The last general chromatographic scheme to be discussed involves the formation of sulfur monoxide (SO) and its subsequent oxidation by ozone to yield electronically excited sulfur dioxide (SO_2^*).[10] Sulfur-selective detection using this reaction depends upon the production of the diradical SO from sulfur-containing analytes; several different methods have been employed to generate SO for this purpose.[11-15]

Chemiluminescence in chromatographic detection can occur from either vibrationally or electronically excited species. In this chapter, vibrationally excited ground state species will be denoted by a superscript dagger (†) and electronically excited species by an asterisk (*). Distinction has been made in the past between "hot" CL, in which photoemission is observed directly from a flame, and "cool" CL, which occurs solely as a result of a chemical reaction.[16] An example of hot CL is the FPD process, and examples of cool CL are analyte reactions with F_2 in the fluorine-induced CL detector. However, such distinction is less clear in the case of $SO + O_3$ chemiluminescence. In this method, the SO is usually formed in a flame, but photoemission is observed under cool, low-pressure conditions.[12] In any case, any chemical reaction that produces photoemission from excited state products is correctly termed chemiluminescent. This usage is meant to distinguish these processes from photoemission from species formed in and excited directly by plasmas or high-energy discharges.

The three schemes for sulfur-selective CL detection, namely, flame photometry, fluorine-induced CL, and ozone-induced CL, and their applications to selective chromatographic detection of sulfur-containing species are discussed below.

1.2. FLAME PHOTOMETRY

1.2.1. Chemiluminescence Mechanism

Historically, the most often used sulfur-selective CL detector in chromatography has been the FPD. Early reports of photoemission from sulfur-containing compounds in reducing (hydrogen-rich) flames involved flame photometry[17-19]; later, Brody and Chaney applied this technology to chromatographic detection[20]. When sulfur-containing molecules are combusted in a hydrogen–oxygen flame, a complex series of radical-induced reactions ensues,[21] producing a variety of sulfur species in excited electronic states. Analogous reactions occur to phosphorus-containing compounds in such flames, resulting in photoemission from HPO* and other excited-state species. In addition to sulfur- and phosphorus-compound detection, FPDs have been applied to the selective detection of other heteroatomic species, notably metal compounds. This discussion will focus on the application of the FPD to sulfur-selective detection in chromatography.

Muller et al. have developed a model using 10 rapid equilibria describing the primary reactions of sulfur in reducing hydrogen–oxygen flames:

$$H_2S + H \rightleftarrows H_2 + SH \tag{1}$$

$$SH + H \rightleftarrows H_2 + S \tag{2}$$

$$SH + S \rightleftarrows H + S_2 \tag{3}$$

$$SH + SH \rightleftarrows H_2S + S \tag{4}$$

$$H_2S + OH \rightleftarrows H_2O + SH \tag{5}$$

$$SO_2 + H \rightleftarrows OH + SO \tag{6}$$

$$SO + H \rightleftarrows OH + S \tag{7}$$

$$SO + H \rightleftarrows SH + O \tag{8}$$

$$O_2 + H \rightleftarrows OH + O \tag{9}$$

$$O + H_2 \rightleftarrows OH + H \tag{10}$$

Both band and continuum emission from sulfur-containing species are observed in FPD flames, and early research studies concluded that the prominent bands at 384 and 394 nm were due to excited-state S_2^*.[19,20] Other bands are due to SO* and SH*,[22] and the continuum has been ascribed to SO_2^*.[23] The number of sulfur species present in such flames, together with emission information, indicates that attribution of sulfur emission in the FPD solely to S_2^* is probably an oversimplification. In general, the sulfur emission in an FPD ranges from 320 to 460 nm, with a maximum at 394 nm. By observation of the flame envelope, sulfur compounds can be detected by their characteristic emission.

Questions still exist about the excitation mechanisms leading to the emitting species in the FPD. For example, several plausible mechanisms have been proposed to explain the formation of S_2^*. One proposal is that the recombination of atomic hydrogen drives the formation of S_2^*.[19,24] Another hypothesized mechanism involves two- or three-body recombination of S atoms in reactions sufficiently exothermic to leave the product, S_2^*, in an excited state.[23,25,26]

1.2.2. Chemiluminescence Efficiency

There currently exists no accurate estimation of the overall quantum yield for chemiluminescence in the FPD. Factors that must be taken into account include the formation, excitation, and relaxation of each of the several CL species thought to contribute to the overall signal. Uncertainties in yields for the first two processes have already been noted, and values for the third process can only be approximated. For example, a typical excited species is estimated to undergo about 200 collisions in an FPD flame over its radiative lifetime,[5] so nonradiative decay of the excited state must be significant.

1.2.3. Detector Design

Despite the ongoing controversy over the characterization of the observed sulfur emission in the FPD, this detector continues to be widely used for sulfur-selective detection in chromatography. Since its inception, many modifications to the original design have been reported in efforts to improve sensitivity, selectivity, and linearity. Most modifications are based on two fundamental FPD designs, incorporating either single- or dual-flame burners.

1.2.3.1. Single-Flame FPD

A schematic of a single-flame FPD is shown in Figure 1.1(a). In this configuration, a photomultiplier tube (PMT) is positioned above—yet shielded from direct viewing of—the lower part of a diffusion flame.[20] Hydrocarbons and other potential interferences emit strongly in the lower part of the flame, whereas sulfur emission occurs slightly above this region. FPD selectivity for sulfur can be maximized by positioning the PMT above the lower flame region and by using a narrow bandpass filter to isolate sulfur emission at 394 nm. Limitations of the single-flame design include flame extinguishment during solvent elution, CL quenching, compound-dependent responses, and interferences, both positive and negative, from hydrocarbons and other species present in the sample.

1.2.3.2. Dual-Flame FPD

The diagram presented in Figure 1.1(b) is of a dual-flame FPD developed to address some of the limitations of the single-flame design.[24] In the

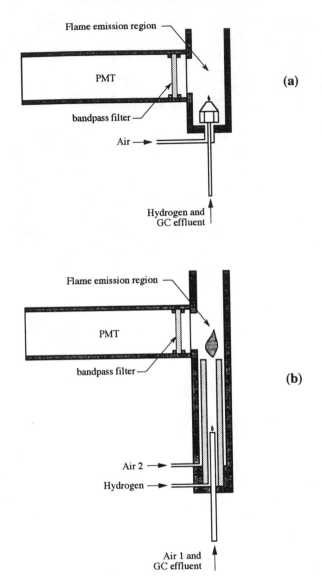

Figure 1. FPD schematics: (a) single-flame design; (b) dual-flame design.

dual-flame design a small, fuel-rich H_2/air flame is positioned before the analytical flame to decompose analyte and matrix compounds eluting from the chromatographic column. Excess unburned H_2 from this first flame is combined with a second supply of air to generate a second flame, where photoemitting sulfur species are formed and excited. A PMT equipped with a filter detects emission above the second flame. In separating sample decomposition from analyte excitation, several limitations of the single-flame FPD can be overcome.[27] Hydrocarbon quenching of sulfur emission can be reduced, and response dependence on analyte structure can be minimized. If the first flame is extinguished by solvent elution, flashback from the larger analytical flame serves to reignite it. At the cost of some sensitivity, the dual-flame FPD design provides a more nearly quadratic response (see the discussion in Section 1.2.4, below) than does the single-flame design.[27] Where hydrocarbons constitute a major component in an sulfur-containing sample, such as in gasoline analysis, the dual-flame design can provide a significant improvement.

1.2.4. Response Characteristics and Applications

The most noted undesirable characteristics of the FPD are its nonlinear response and its susceptibility to interferences. Experimentally derived response curves for the instrument follow the general form,

$$\text{response} \propto [S]^n$$

where n ranges from 1.6 to 2.4 in modern designs.[28] From the stoichiometry of the S atom recombination reaction alone, one would expect $n = 2$, where a 10-fold decrease in S atom concentration would lead to a 100-fold decrease in signal. This characteristic leads to poor detection at low analyte concentrations. A variety of explanations for noninteger n values have been proposed and are thoroughly reviewed elsewhere.[5]

It is common to show calibration curves for the FPD on log–log graphs, in an attempt to linearize the signal. A dynamic range of 3 orders of magnitude has been reported for some FPD designs.[29] Dynamic range is limited by another salient feature of FPD calibration curves, termed the saturation region. This region is characterized by a loss of sensitivity at higher analyte concentrations. Again, several theories have been advanced to explain this observation. There is some consensus on two theories, namely, collisional quenching of excited states by flame species such as H_2O and SO_2, and loss of the precursor radicals H and OH via three-body recombination effected by SO_2 or organic radical scavengers.[24,27,29-31] Linearizers using electronic circuitry are available for many FPDs, but these assume $n = 2$ and may contribute to quantitation errors.[5] Linearization of response to sulfur has been attempted by doping the flame with small amounts of a sulfur compound, typically SO_2 or SF_6.[32] This technique requires extreme care to ensure a steady supply of the sulfur-containing dopant. A major drawback is the loss of dynamic range, which is reduced to only 1–1.5 orders of magnitude when dopant is added.[5]

Detection limits reported for FPDs are dependent on several factors including sulfur compound investigated, detector manufacturer and design, and flame gas composition.[33] Typical reported detection limits range from 2 to 350 pg S/s for simple sulfur gases such as H_2S, SO_2, and COS.[29] Selectivity is also dependent on several factors and typical response values range from 10^4 to 10^6 vs. the signal from hydrocarbons.

Since its development in 1966, the FPD has been used extensively with gas chromatography (GC). More recently, sulfur-selective detection with the FPD has been reported in analytical liquid chromatography (LC),[34] microbore LC,[35] ion chromatography,[36] and supercritical fluid chromatography (SFC).[37,38] Because of the long history of FPD use in sulfur-selective chromatographic detection, a wide variety of applications of this detection scheme have been reported and are too numerous to mention here. The interested reader is referred to recent review articles on the FPD for application information.[5,39] The FPD is increasingly being replaced by the sulfur chemiluminescence detector based on the $SO + O_3$ reaction (see Section 1.4, below), which is more sensitive, has a wider linear range, and does not suffer as much from interferences.

1.3. FLUORINE-INDUCED CHEMILUMINESCENCE

1.3.1. Chemiluminescence Mechanism

The second chromatographic CL scheme discussed in this chapter utilizes the reaction of reduced sulfur compounds with molecular fluorine to produce chemiluminescence.[6] When certain sulfur compounds react with F_2 at reduced pressure ($\leqslant 2$ torr) in the gas phase, rich chemiluminescence from a variety of emitting species is observed. Conclusions drawn from spectral studies of these reactions have differentiated the sulfur-containing analytes into two general groups: those producing mainly $HF^\dagger$ and those producing mainly other species.[8] The former group of compounds, including the longer chain thiols, sulfides, and disulfides, exhibits spectra characteristic of vibrationally excited HF, mostly with $\Delta v = 4$. The latter group produces emission from electronically excited CH_2S^*, HCF^*, and possibly FCS^*; this group includes the methyl-substituted species CH_3SH, dimethyl sulfide (DMS), and dimethyl disulfide (DMDS).[8] The effect of the presence of typical high-performance liquid chromatography (HPLC) solvent vapors on chemiluminescence was also investigated. Methanol diminishes emission from $HF^\dagger$ bands while enhancing an unassigned, broad spectral emission feature; acetonitrile has relatively little effect on any $HF^\dagger$ features in the observed spectra. CH_2S^* features, normally weak or absent in the spectra of the longer chain mercaptans, appear strongly in the presence of methanol.

Glinski et al.[8,40] proposed mechanisms for the formation of the emitting species. Molecular fluorine, rather than atomic fluorine in equilibrium with F_2, was proposed to be the initial species to attack the sulfur compound in these reactions. The general reaction

$$R_1R_2S + F_2 \rightarrow R_1R_2SF + F \tag{11}$$

is considered the rate-limiting step leading to chemiluminescence and is the route by which F atoms are produced (8). $HF^\dagger$ is then produced by either H-atom abstraction.

$$F + RSCH_3 \rightarrow HF^\dagger + RSCH_2 \tag{12}$$

or by hydrogen atoms, produced in an intermediate reaction with S-compound analyte, reacting with molecular fluorine,

$$H + F_2 \rightarrow HF^\dagger + F. \tag{13}$$

The mechanism of formation of CH_2S^* is thought to be slightly different.[40] Finally, HCF^* formation is not well understood. The most energetic emission from this molecule is at 480 nm, corresponding to at least 65 kcal/mol of energy.[8]

When atomic fluorine is used directly as the reagent, sulfur selectivity is lost and CL response becomes nearly universal.[4] Because the reactions of analytes with atomic fluorine are not selective, they will not be discussed here.

1.3.2. Chemiluminescence Efficiency

Efficiency in the F_2-induced CL design is determined in large part by reaction rates of sulfur-containing species with F_2. In addition, the reactants are mixed and swept out of the cell on a timescale permitting only a fraction, less than 1%, of the fluorine molecules to react. At the low pressure of the reaction cell, however, an appreciable (in comparison to the FPD) number of excited state products can be expected to chemiluminesce. Although no overall number was obtained for the CL quantum yield, based on the low detection limits for reduced sulfur compounds (see below) this scheme can be considered relatively efficient.

1.3.3. Detector Design

In the F_2-induced CL detector, chromatographic effluents are directly mixed with fluorine in a CL reaction cell maintained at low pressure. Heating tape is used to prevent cold spots in the transfer line from the GC to the detector. Fluorine is either supplied as a mixture of 5% F_2 in He or produced upstream of the reaction chamber by a high-frequency discharge in SF_6. The latter method circumvents the health and safety problems associated with storage and handling of fluorine gas mixtures. Residence time of analyte and reagent in the reaction chamber is approximately 100 ms. A cooled, red-sensitive PMT is used to observe emission from sulfur compounds and is fitted with a bandpass filter to isolate the spectral region of interest, from 660 to 740 nm. Exhaust from the cell passes through a chemical trap to destroy unreacted F_2, HF, and other products before entering the vacuum pump. A diagram of the F_2CL cell is shown in Figure 1.2.

1.3.4. Response Characteristics and Applications

The F_2-induced CL detector responds to a range of thiols, sulfides, and disulfides; thiophenes, however, respond a factor of 10 less strongly, presumably owing to the relative stability of the aromatic ring. The sulfur-containing gases SO_2, COS, H_2S, and CS_2 are not sensed by this detector.

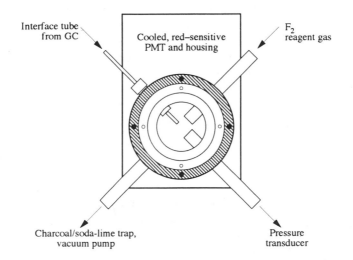

Figure 1.2. F_2-induced chemiluminescence cell schematic. From Mishalanie and Birks,[7] with permission.

Response of the F_2 CL detector is linear over 4 orders of magnitude, from 100 pg to 1 μg in DMS concentration. Detection limits, which are compound dependent, range from 5 to 1800 pg.[6,7] Thiols and sulfides are detected in the low picogram range, and responses from all compounds are dependent on the reagent gas flow rate.

Sulfur selectivity in the F_2-induced CL scheme over saturated hydrocarbons and many oxygen-containing compounds is excellent; measured selectivity vs. hexane is $> 10^7$.[6] However, a variety of compounds that do not contain sulfur also exhibit a CL response upon reaction with F_2. Examples of interfering species include alkenes, aldehydes, compounds with weak C—H bonds such as halocarbons, toluene, and xylene,[16] reduced heteroatomic species like phosphines and arsines,[41] and methylated tellurium and selenium compounds.[42] The lack of selectivity in detecting sulfur compounds in the presence of aromatic and olefinic compounds effectively excludes the application of F_2-induced CL detection from analysis of many complex petrochemical matrices in which sulfur compounds are found.

Despite this significant lack of selectivity, F_2 CL detection has been used

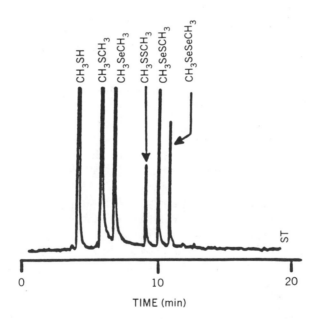

Figure 1.3. Headspace analysis of volatile sulfur and selenium compounds from a culture of a bacterium, *Pseudomonas fluorescens* K27, analyzed by the F_2-induced CL detector. Separation accomplished on a 30 m DB-5 capillary column of 0.25 mm i.d. and 1.0 μm stationary phase thickness. The GC temperature program was $-20\,°C$ for 1 min, -20 to $300\,°C$ at $20\,°C/min$, and held at $300\,°C$ for 5 min. From Chasteen et al.,[42] with permission.

successfully in GC,[6,41,42] SFC,[9] and microbore LC.[7] Analyses of a variety of matrices for sulfur compounds have been reported, including sulfides and thiols in natural gas,[43] beer,[7] bacterial and fungal headspace samples (see Figure 1.3),[42] and the pesticide malathion in a commercial insecticide formulation.[7,9]

1.4. OZONE-INDUCED CHEMILUMINESCENCE

1.4.1. Introduction

In 1966, a series of reports described the chemiluminescent reactions of sulfur monoxide (SO).[44-46] When SO is oxidized by ozone, electronically excited SO_2 molecules are formed owing to the high exothermicity (106 kcal/mol) of the reaction. SO_2^* then relaxes by emission of a photon:

$$SO + O_3 \rightarrow SO_2^* + O_2 \qquad (14)$$

$$SO_2^* \rightarrow SO_2 + h\nu \qquad (15)$$

Luminescence and phosphorescence are both observed; emission occurs from approximately 280 to 450 nm and is most intense near 350 nm, with both continuous and banded features. This region is relatively free of spectral interferences, and most PMTs are quite sensitive at these wavelengths. SO_2^* emmission is also energetically efficient; the highest frequency emission observed corresponds to an energy of 101 kcal/mol, or 95% of the total energy available.[10] The detector in which SO is allowed to react with O_3 to form SO_2^* is known as the sulfur chemiluminescence detector (SCD®).[47] The use of this reaction in chromatographic detection has afforded extremely sensitive and selective methods for sulfur analysis, and to date the SCD appears to be the detector of choice for such analyses. It is simpler and less expensive than an atomic emission detector (AED) and it is more sensitive and selective than the FPD or other sulfur-selective detectors.

The challenge of implementing ozone-induced chemiluminescence for chromatographic detection has been to form SO effectively prior to its reaction with ozone in Equation (14). Calculations have predicted the lifetime of SO to be less than 2 ms in air at room temperature and 1 torr pressure.[48] However, it now appears that SO is relatively stable in both the gas phase and the liquid phase in the absence of oxygen, and it survives transport through membranes before undergoing reaction with ozone. Despite the purported instability of SO, several methods for forming and transporting

SO to a CL reaction chamber have been published. For the purpose of this chapter, chromatographic detection schemes based on the SO + O_3 reaction will be divided into several types, based on the process by which the diradical species, SO, is formed. Processes of SO production for chromatographic sulfur-selective CL detection include the following: direct reaction with ozone[11]; flame conversion, termed flame SCD[49]; flameless combustion, or flameless SCD[13]; catalytic thermionic conversion, or sulfur nitrogen–phosphorus detection[14]; and aqueous phase oxidation.[15] These methods are discussed below.

1.4.2. Direction Reaction with Ozone

1.4.2.1. Chemiluminescence Mechanism

When ozone is mixed with certain reduced sulfur compounds, a complex series of radical chain reactions occurs and emission from electronically excited SO_2 is observed.[50] H_2S, mercaptans, sulfides, sulfoxides, and thiophenic compounds produce weak CL upon direct reaction with ozone. CL spectra from these reactions have been identified as arising from SO_2^*.[51–54] Formation mechanisms are pressure and compound dependent and are not well established. In general, ozone is thought to attack a C—S bond,

$$CH_3SH + O_3 \rightarrow H_3COO + HSO \tag{16}$$

proceeding through a cyclic Criegee-like intermediate.[51,55] Reaction with excess ozone produces sulfur monoxide,

$$HSO + O_3 \rightarrow OH + SO + O_2 \tag{17}$$

The SO formed in this reaction is subsequently oxidized via the reaction cited in Equation (14) to produce SO_2^*. The reactions shown here are an example of a proposed mercaptan oxidation pathway; those proposed for H_2S,[50,55,56] sulfides,[50,54,55,57] sulfoxides,[54] and thiophenes[53] are qualitatively similar in that several ozone molecules are required to oxidize each sulfur-containing molecule. At extremely low (millitorr) pressures, single-collision pathways to SO_2^* are possible. Weak emission from electronically excited HSO* has been observed below 100 mtorr[54,55]; however, the chromatographic detection schemes described here operate at much higher pressures (10–20 torr). In this pressure range, multiple collisions are the rule and are much more efficient at producing chemiluminescence; therefore, this discussion is confined to multistep pathways to SO_2^*.

1.4.2.2. Chemiluminescence Efficiency

With the possible exception of the flame conversion process, none of the $SO + O_3$ CL schemes are understood well enough to make quantitative assessments of efficiencies. Any such estimate, therefore, is ultimately based on the reported detection limit or sensitivity toward various sulfur-containing compounds detected in a particular reactor paired with a particular detector design. For example, direct reaction of reduced sulfur compounds with ozone exhibits widely varying sensitivities dependent on the sulfur compound reacted. Sensitivity is determined by the functionality of the S atom in the analyte, with response decreasing in the order $CH_3SH > CH_3SCH_3 > H_2S >$ thiophene.[56] The rate of initial ozone attack on the analyte molecule seems to determine sensitivity, and therefore efficiency, in this detection scheme. Oxidized sulfur compounds such as SO_2 must first be reduced to SO; thus, these compounds are not detected by direct reaction with ozone.

1.4.2.3. Detector Design

Detect reaction with ozone requires no pretreatment of a sulfur-containing analyte. Gas chromatographic effluent is directed into a CL reaction chamber at reduced pressure and mixed with ozone directly in front of a PMT; the reaction chamber design[58] is essentially identical to that shown in Figure 1.2 for the F_2-induced CL detector. Use of this scheme in LC requires an enclosed interface to collect and dispose of the liquid waste.[11] LC effluent is nebulized by a high-velocity stream of ozone in air, which also serves as the reagent gas. The fine aerosol spray produced in the nebulization process collects on the inside of the glass reaction cell and is drained off to waste. This design minimizes memory effects.

1.4.2.4. Response Characteristics and Applications

This scheme has been termed reduced sulfur chemiluminescence, indicating the lack of response to oxidized sulfur compounds. Response factors are highly structure dependent for the sulfur compounds detected by direct reaction with ozone. Olefins also give rise to a CL signal, with ethylene responding most strongly; however, 3 orders of magnitude selectivity for detection of sulfur compounds over the response of olefins of higher molecular weight than ethylene was reported.[58] Other non-sulfur-containing compounds that give a measurable CL response include polyaromatics, nitrogen heterocycles, hydrazines, azide dyes, and other fluorescing molecules.[11] Sulfur selectivity was reported as $> 10^7$ over hydrocarbons.[58]

Sulfur detection by direct reaction with ozone has been applied to GC[58,59] and LC.[11] Reported detection limits ranged from 2 ng for DMS to 250,000 ng for phenyl sulfide. Sulfur response was linear over 3 orders of magnitude, with no reported interference from a coeluting hydrocarbon solvent.

It should be emphasized, however, that sulfur CL emission is orders of magnifude weaker (expressed as relative molar response factor) than that observed when sulfur compounds are first converted to SO prior to oxidation by ozone. Therefore, direct reaction of sulfur analytes with ozone has not been analytically practical; in fact, direct reaction is insignificant compared to flame conversion with respect to CL signal generation. Sulfur-selective chromatographic detection by direct reaction with ozone has not found wide application, perhaps due to the weak and varying detector responses and the number of non-sulfur-containing interferences exhibited by this scheme.

1.4.3. Flame Conversion

1.4.3.1. Chemiluminescence Mechanism

Sulfur monoxide can be formed efficiently when sulfur compounds undergo combustion. In reducing flames, SO is the second most prevalent sulfur species, approaching 15 mol%,[21,23] quantitatively second only to SO_2. By directing chromatographic effluents into a hydrogen-rich H_2/air flame, and by sampling the gaseous products of the flame into a CL reaction chamber at reduced (< 20 torr) pressure, one can achieve sulfur-selective CL detection of a wide variety of sulfur-containing compounds.[12,49]

Unlike direct reaction of a sulfur-containing analyte with ozone, oxidized as well as reduced species are detected downstream of the flame with nearly equimolar response. Others have suggested that sulfur compounds in the flame, regardless of original structure, are decomposed to sulfur atoms in nearly identical yields.[60] Sulfur atoms then react according to the general pathway

$$S \rightarrow SO \rightarrow SO_2 \qquad (18)$$

as they propagate through the flame. A hollow ceramic sampling probe, made of high-purity alumina, is inserted into the region of maximum SO concentration and used to withdraw most of the flame gases, including SO, under vacuum and rapidly (~ 1 s) transport them to a CL reaction chamber. The flame ionization detector response can be quantitated along with the CL emission response to provide simultaneous general and sulfur-selective chromatograms of a sample.[61] Table 1.1 compares the detection limits of the FPD, the F_2-induced CL detector, and the flame SCD for a variety of sulfur compounds.

Table 1.1. Intercomparison of Sensitivities of Detection of Sulfur Compounds in Air by Chemiluminescence Detectors[a]

	FPD	F$_2$-Induced CL	Flame SCD
H$_2$S	130	Not detected	5
COS	90	Not detected	3
CH$_3$SH	240	7	3
C$_2$H$_5$SH	290	4	3
CH$_3$SCH$_3$	120	3	2
CH$_3$SSCH$_3$	180	1	1

Source: Data obtained by R. Dominguez, South Coast Air Quality District Laboratory, Los Angeles. Presented at the Symposium on Selective Detectors, American Chemical Society meeting, Washington, DC, August 1992.

[a]Lower limit of detectability, expressed as parts per billion (10^{-9}).

The flame SCD is shown to be the most sensitive of the three; additionally, this detector does not suffer from the interference problems of the other two schemes in Table 1.1 (see below).

1.4.3.2. Chemiluminescence Efficiency

The efficiency of flame conversion of sulfur compounds to SO has been determined by gas phase titration with NO$_2$.[62] SO reacts rapidly with NO$_2$ in the gas phase according to Equation (19)[63]:

$$SO + NO_2 \rightarrow SO_{2(\text{ground state})} + NO \qquad (19)$$

Addition of NO$_2$ should therefore scavenge any SO in the permeant gas stream; consequently, the CL signal should decrease with added NO$_2$. The flame yield of SO from SO$_2$ was estimated to be approximately 0.4% by this method. The difference from the thermodynamically calculated yield was ascribed to perturbation of the flame by the sampling probe. An overall efficiency for this method of one detected photon per 10^9 sulfur molecules entering the flame was calculated; for comparison, approximately one CH$_2$ group in 10^6 produces a detectable ion in a flame ionization detector.

1.4.3.3. Detector Design

A diagram of the flame SCD design is shown in Figure 1.4. A hydrogen-rich flame is used to convert analyte sulfur compounds to SO. One major advantage of this design is that a flame ionization detector, standard on most

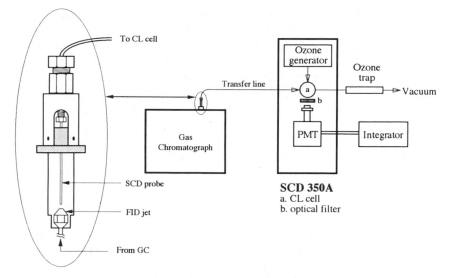

Figure 1.4. Flame SCD schematic.

GC instruments, is used to generate the flame. In early studies, flame gas flow rates were quite different than in normal FID operation. For example, typical flows for flame SCD operation were 200 mL/min H_2 and 400 mL/min air. More recent work has utilized flow rates more typical of traditional FID operation. The FID signal provides valuable additional information about non-sulfur-containing species at nearly the same sensitivity, even with the sampling probe inserted in the flame. The added dimension provided by generating simultaneous general and sulfur-selective chromatograms can greatly facilitate sample characterization.[61]

Flame gases are sampled by inserting a 0.5 mm i.d. ceramic probe into the flame and withdrawing effluent gases under vacuum. It has been shown that dropping the pressure by a factor of 10 or more rapidly cools the flame effluent and can quench chemical reactions,[64] preventing further oxidation of SO to the more thermodynamically stable SO_2. The pressure drop is effected along the length of the sampling probe. The pressure downstream of the probe is held below 20 torr to prevent condensation of water (produced in large amounts by the flame) in the transfer line or CL chamber. Up to 20 ft of opaque poly(tetrafluoroethylene) (PTFE) transfer line has been used to connect the sampling probe with the CL chamber with no appreciable loss of sensitivity. A stream of air is passed through an electrical discharge to produce ozone before being mixed with the flame effluent in the CL chamber. Chemiluminescence is monitored by a PMT equipped with a filter to select emission in the blue while excluding emission at different wavelengths. Because oxides of nitrogen are formed in the flame, the filter is necessary to discriminate against the light emitted at >600 nm by the $NO + O_3$ reaction. A chemical trap downstream of the CL chamber destroys excess ozone and other reactive gases before they reach the vacuum pump.

1.4.3.4. Response Characteristics and Applications

Because combustion in a flame converts all sulfur-containing analytes to SO in approximately equal yields, this analytical scheme shows nearly equimolar response to sulfur compounds regardless of oxidation state. Thus, an estimate of the total sulfur content of a sample can be obtained without having to identify individual species. This capability is useful in fuel analysis, where sulfur content can be critical in determining the quality of a fuel. Furthermore, through adjustment of the residence time of the sample within the flame, nearly complete selectivity over non-sulfur-containing compounds can be achieved. Under these conditions, no interference from coeluting compounds is noted.[65] Selectivity over hydrocarbons is greater than 10^7.[49] Flame conversion shows a linear range of 5 orders of magnitude, and detection limits have been reported as low as 0.4 pg S/s when the conditions were fully optimized.

The combination of high sensitivity and selectivity, wide linear range, and lack of quenching from non-sulfur species has made flame SCD a popular choice for chromatographic detection of sulfur. This design has been applied to GC,[49,65-69] SFC,[70] and capillary LC.[71] The use of the flame SCD and its coupling to GC, SFC, and capillary LC has been recently reviewed.[72] Analyses of sulfur gases,[28,68,73] phenylthiohydantoin(PTH)–amino acids,[71] n-alkylsulfonic acids,[71] sulfur-containing pesticides,[65,70,71,74] polycyclic aromatic sulfur-containing hydrocarbons,[74] nonionic surfactants[70] and sulfur compounds in beer,[69] natural gas,[66] diesel fuel,[74] and petroleum streams[49,66-68,75] have been reported. The SCD was found to be more sensitive, show higher long-term stability, and be less dependent on fluctuations in atmospheric pressure than an FPD.[28] The SCD was also compared to an atomic emission detector (AED)[75] and found to be comparable for selective sulfur analysis in most respects, at about one-fifth the cost.

The flame SCD does have some limitations, however, due to use of a flame to generate SO. CL response in this scheme is highly dependent on several factors, including H_2 and air flows and sampling probe surface conditioning. Response is especially dependent on the radial and axial position of the probe within the flame. There is some chemical noise produced by the flame, possibly due to emission from OH and/or NO reactions with ozone in the CL cell. The flame produces a large amount of water vapor, requiring frequent maintenance of the traps and vacuum pump. These characteristics necessitate careful attention to proper operation and maintenance to get the best performance from the flame SCD.

1.4.3.5. Flameless Combustion

Although generation of SO in a flame has become a widely used sensitive and selective method of sulfur detection, there are several advantages realized by the use of other ways to produce SO from sulfur-containing compounds. Recently a flameless combustor to convert sulfur compounds to SO for subsequent SO + O_3CL detection has been described.[13,76] This combustor is reviewed extensively in Chapter 2 of this book.

1.4.4. Catalytic Thermionic Conversion

1.4.4.1. Method of SO Formation

In another new method of SO formation, we have utilized the hot thermionic source of a nitrogen–phosphorus detector (NPD) to convert sulfur compounds eluting from a chromatographic column into oxides of sulfur, including SO.[14] The NPD source used is an externally heated rubidium-doped silicate bead in a H_2/air atmosphere; surface temperatures are roughly 800–1100 °C. The chemistry of the conversion process is not well understood, but an analogy may be drawn from proposed nitrogen and phosphorus chemistry at NPD beads.[77] Sulfur compounds are first adsorbed onto the surface of the bead, presumably by lone pair interaction with a surface site. Conversion of the sulfur compound to SO occurs; the SO can be withdrawn under vacuum by a probe inserted into the NPD source region and allowed to react with ozone to form SO_2^*. No flame is employed in this detector. SO production is inferred by enhanced CL response to dimethyl sulfoxide (DMSO) when the bead is heated; no response to DMSO is observed with a cool, inactive bead.

1.4.4.2. Chemiluminescence Efficiency

The efficiency of SO formation by catalytic thermionic conversion of sulfur-containing compounds is unknown. Based on reported detection limits, it is certainly less efficient than the flame conversion process; however, this method still exhibits sub-ppm detection limits for reduced sulfur compounds.

1.4.4.3. Detector Design

This design utilizes the heated bead of a standard Hewlett Packard NPD detector to convert sulfur compounds to a readily detectable species (see

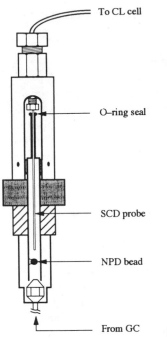

To CL cell

O–ring seal

SCD probe

NPD bead

From GC   **Figure 1.5.** SNPD schematic.

Figure 1.5). Gas flows are identical to those used for normal NPD operation, ca. $4\,mL/min$ H_2 and $50\,mL/min$ air. A ceramic probe is inserted into the thermionic source region to within a few millimeters of the bead. Flow through the probe creates a partial vacuum in the thermionic chamber, necessitating a higher-then-normal electric current to heat the bead to operating temperature.[78] Gaseous products of bead surface chemistry are withdrawn under vacuum and directed into the CL chamber maintained at reduced pressure.

The presence of the probe perturbs the normal NPD chemistry, as shown by a 10-fold decrease in sensitivity for nitrogen and phosphorus compounds and a higher heating current required to sustain surface reactions.[14] However, the NPD signal still provides selective and sensitive detection for N and P compounds. The SNPD allows two chromatograms to be obtained simultaneously, one providing information on sulfur-containing analytes present in the sample, and the other on nitrogen- and phosphorus-containing analytes also present.

1.4.4.4. Response Characteristics and Applications

When the NPD detector is operated normally, with the exception of using slightly higher bead heating current, sensitive and selective sulfur detection by CL is obtained without a large decrease in NPD performance. In contrast to the flame SCD design, this scheme is relatively simple to optimize, showing much less dependence on detector gas flows and probe position. Exact positioning of the probe (critical for optimum sensitivity in a flame) is not required, and the probe tip can be anywhere from 2 to 20 mm above the bead without marked changes in response. In addition, the SNPD operates with gas flow rates lower than those of the flame SCD, leading to decreased water formation by about an order of magnitude. These differences translate into prolonged chemical trap lifetimes and suggest possible application as a process detector.

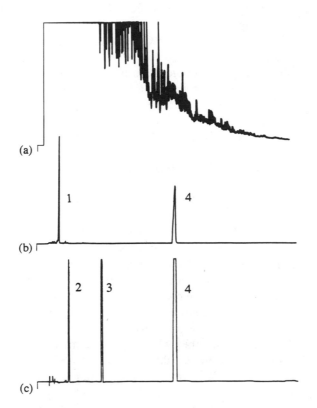

Figure 1.6. Selective two-channel operation of the SNPD (b & c), with comparison to detection by FID (a). Only those compounds containing S, N, or P are detected. These two chromatograms result from a single injection of a mixture of SNP compounds diluted in a complex mixture of hydrocarbons. Conditions were as follows: 1.0 μL of a solution of about 100 ppm of each S, N, and P compound and 10% (v/v) gasoline in hexanes was injected, split 1:40; GC oven temperature program: 80 °C for 2 min, ramped at 8 °C/min to 160 °C and held at 160 °C for 2 min. Bead heating voltage 22.45 V. (1) DMSO; (2) trimethyl phosphate; (3) 4-t-butylpyridine; (4) dimethylthiourea.

Use of the SNPD in GC detection has shown linear response to sulfur compounds over 3 orders of magnitude. The detection limit for sulfur is 1.5 pg S/s, about twice that of the flame SCD in the same study.[14] No measurable quenching of sulfur CL response is observed when up to 10% (wt/wt of sample injected) of hydrocarbons coelute with the sulfur-containing analyte. Greater amounts ($>10\%$ wt/wt) of hydrocarbon begin to interfere with CL response by quenching the reactions at the bead surface. The SNPD design permits simultaneous sulfur and nitrogen–phosphorus compound detection without sample splitting (see Figure 1.6), shows high selectivity, is easily optimized, exhibits low gas consumption, and does not require pre-conditioning of the ceramic sampling probes. However, although all reduced sulfur compounds tested were detected by the SNPD, of those tested only two compounds with sulfur in higher oxidation states, DMSO and thiirane-1-oxide, gave a response; SO_2, SF_6, dimethyl sulfate, and tetramethylene sulfone were not detected. It should be noted that these compounds are, however, readily detected by the flame SCD.

1.4.5. Aqueous Phase Oxidation

1.4.5.1. Mechanism of SO Formation

We have recently developed a sulfur-selective detector for liquid chromatography. The aqueous-phase reactions involved appear to be different from those observed in gas-phase detectors. We have obtained evidence for SO formation in the liquid phase by heating sulfur compounds dissolved in aqueous solutions at 200–300 °C under pressure sufficient to maintain the sample in the liquid phase.[15] The detector apparatus consists of an uncoated, heated capillary tube reactor, followed by a cooled flow restrictor leading to a permeation cell maintained at atmospheric pressure. In the permeation cell, the aqueous solution is directed to waste, while the permeant gases are swept into a CL reaction chamber where they are mixed with ozone. This detector is called the liquid-phase sulfur analyzer (LPSA).

The earlier methods for SO formation are similar in that the conversion reactions producing SO all take place in the gas phase. When aqueous sulfur compounds are swept through a heated metal or fused silica reaction capillary and product gases are allowed to permeate across a hydrophobic, microporous membrane into the gas phase, sulfur-selective CL emission is observed in the gas phase upon oxidation of the product permeant gases by ozone. No flame is used in this system. Addition of oxidizing or reducing agents to the aqueous stream produces noticeable effects on the conversion, but sensitive CL detection of aqueous sulfur compounds is also observed with no added reagent except the water used as a mobile phase, which may also contain trace concentrations of dissolved oxygen.

Candidates for the CL-active species produced in this scheme must contain sulfur, be relatively volatile, have a reasonably large Henry's law constant, be produced in the oxidizing conditions in the reaction capillary, and be relatively long lived (seconds) in aqueous phase. H_2S has been shown to be efficiently destroyed in the heated reaction zone and produces no detectable signal,[15] so it can be ruled out as the CL-active intermediate. Other possible CL-active species are volatile organic sulfur compounds such as DMS or CH_3SH. However, mass spectrometric experiments have detected no H_2S, DMS, or CH_3SH in the permeant gas stream. Since H_2S is not detected and is not formed from other analyte species in this detector, SO appears to be the species formed and detected.

Additional experiments to confirm SO formation from sulfur compounds by aqueous phase oxidation have substantiated the evidence provided by mass spectrometry. A known source of SO is the thermal decomposition of thiirane-1-oxide, according to Equation (20) below[79]:

$$\underset{\underset{H_2C-CH_2}{\overset{\diagup\ \diagdown}{\underset{}{}}}}{\overset{\overset{O}{\overset{\uparrow}{}}}{S}} \xrightarrow{100\,°C} SO + CH_2{=}CH_2 \tag{20}$$

Repetitive injections of thiirane-1-oxide solutions produced strong CL signals after aqueous phase oxidation, evidence that SO is sufficiently long lived in aqueous solution to be the intermediate in the new detector. The other product of the reaction, ethylene, gives rise to a very weak CL signal upon oxidation with ozone, which cannot account for the strong signal detected in these experiments.

Conclusive evidence for SO being the CL species produced by aqueous phase oxidation of liquid phase sulfur compounds was provided by titration of the permeant gas stream with NO_2.[63] A marked decrease ($>90\%$) in the CL signal from injections of 10-μM DMTU was observed when trace amounts of NO_2 were added to the sweep gas stream of the new detector.

Mass spectrometry experiments, along with results from thiirane-1-oxide injections and gas phase NO_2 titration, have provided evidence for SO as the species formed and detected by aqueous phase oxidation followed by ozone-induced CL. These findings imply considerably greater stability in the aqueous phase for SO than has previously been recognized (see below).

1.4.5.2. Chemiluminescence Efficiency

No quantitative measurement has been made of the conversion efficiency of aqueous sulfur-containing compounds to SO in this detector. Based on relative detection limits, this scheme is estimated to be less efficient than the flame SCD scheme by roughly an order of magnitude.

1.4.5.3. Detector Design

This detection scheme is different from the other $SO + O_3$ schemes in that the conversion process forming SO is carried out in the liquid phase (see Figure 1.7). A high-pressure LC pump establishes liquid flow at 1.0 mL/min to elute samples injected onto a chromatographic column. The liquid stream is then pumped through a reaction capillary tube maintained at 300 °C by resistive heating and 2000 psig by the use of an in-line backpressure valve. Operating the heated reaction capillary tube at elevated pressure is required to prevent vaporization and deposition of analyte and LC mobile phase buffer in the capillary tubing, which can lead to memory effects and plugging. After the reactor, the eluent stream, still maintained under pressure, is passed through a cooling bath and then through a backpressure valve, where the pressure and temperature are allowed to drop to ambient atmospheric conditions. The countercurrent extraction cell then allows any dissolved gases present, including SO, to permeate across a semipermeable membrane into a flowing stream of helium. The SCD vacuum pump draws permeant gases in the helium stream off to the CL cell, where they are mixed with ozone and the SO allowed to react in front of a PMT.

Stainless steel, aluminum, nickel, fused-silica-lined stainless steel, and ceramic capillary tubes have all been used to effect this conversion, with stainless steel appearing to be the most practical and durable material of those investigated. For application to reversed-phase LC, a hydrophobic PTFE membrane is used in the extraction cell. This cell is maintained at atmospheric pressure to prevent failure of the thin membrane. A pressure drop in the gas phase downstream of the PTFE membrane is maintained by a restrictor in the sweep gas exit fitting. The transfer line and CL reaction cell are maintained at about 10 torr with the vacuum pump of the SCD to prevent condensation of water vapor in the gas stream.

1.4.5.4. Response Characteristics and Applications

Aqueous phase conversion has been used to detect many classes of sulfur-containing compounds dissolved in the liquid phase.[15] Representative mercaptans, sulfides, disulfides, sulfones, and sulfoxides, as well as sulfite and

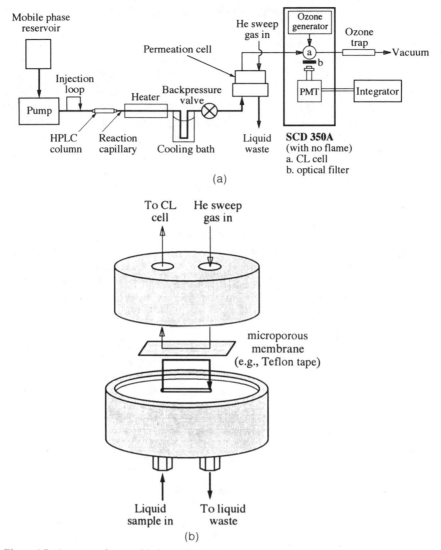

Figure 1.7. Aqueous-phase oxidation schematics: (a) apparatus; (b) countercurrent membrane extraction cell.

sulfate, have all produced a CL signal in this detector. Biologically active sulfur-containing compounds detected by aqueous phase conversion include proteins, peptides, amino acids, pesticides, and pharmaceuticals. Sulfur-containing compounds that produce weak or no CL are sulfate and persulfate salts. Tetramethylene sulfone and representative sulfonic acids gave no detectable response under the conditions tested. In addition, hydrogen sulfide is not detectable in this system.

The most important adjustable parameters in aqueous phase oxidation are the temperature and pressure of the liquid phase during conversion. Many sulfur compounds show a maximum in their CL response when the liquid phase reaction capillary is heated to 300 °C. Higher temperatures presumably convert some of the SO to SO_2, which does not chemiluminesce upon reaction with ozone. Sensitive detection is obtained for proteins and thiocarbamate pesticides at 300 °C, although increased response with increasing temperature is noted. The reaction capillary backpressure is chosen to prevent vaporization of the aqueous solution; the vapor pressure of water at 300 °C dictates that at least 1250 psi be maintained to keep the reaction mixtures in the liquid phase. Operation at higher pressures may stress the LC column and reduce its usable lifetime, but the combination of 300 °C and 2000 psig has been shown to be optimum for the reaction system. The mobile phase pH can be significant for response sensitivity to some sulfur compounds. Surface effects in the reaction capillary have been shown to produce long-term response drift, perhaps due to changes in surface characteristics. These changes can be minimized by injecting a concentrated standard of a sulfur-containing compound prior to analysis. Mixtures of up to 80% organic mobile phase modifier can be accommodated without wetting the hydrophobic membrane in the extraction cell; this limit will accommodate almost all reversed-phase separations. Acetonitrile and methanol have both been used successfully as mobile phase modifiers. The presence of methanol diminishes the CL response; acetonitrile has little effect on detector sensitivity in concentrations up to 75% in water. Once formed, however, SO has been shown to be relatively stable in the liquid stream. The loss of SO in helium-sparged distilled water follows first-order kinetics, and we have calculated a half-life of 0.65 s for SO in aqueous solution in stopped-flow experiments using this system (see Figure 1.8).

Linearity of this detection scheme extends over 3 orders of magnitude; selectivities over most common aqueous interferences (e.g., non-sulfur-containing amino acids and inorganic salts such as chloride and nitrate) and olefinic compounds are greater than 10^6. Two non-sulfur-containing interferences, nitrite ion (NO_2^-) and low-molecular-weight (C_3-to-C_5) alcohols, produce CL signals. Nitrite is converted to nitric oxide in the reaction capillary,[80] and small alcohols are dehydrated to form volatile olefins.[81] Both NO and volatile olefins can permeate the extraction membrane and are known to chemiluminesce upon gas phase oxidation by ozone, although at different wavelengths than does SO_2^*. Although longer chain alcohols may also be dehydrated to form olefins, they do not give rise to an interfering CL response, presumably during to lower permeation efficiencies. Methanol

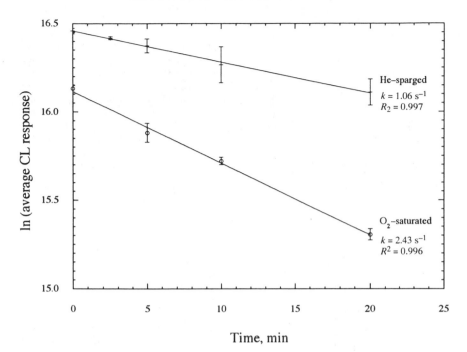

Figure 1.8. Loss of SO in the liquid phase follows first-order kinetics, as shown in this plot of data obtained in stopped-flow experiments in the aqueous phase detector. Liquid flow was stopped 5 s after injection, leaving the analyte band in the PEEK tubing transfer line between the backpressure valve and the permeation cell. Flow was restarted after the indicated time had elapsed. The upper curve was obtained by using helium-sparged doubly distilled deionized (DDI) water as mobile phase and reaction medium, and the lower by using oxygen-saturated DDI water in like fashion. In both cases, loss of SO as measured by CL emission followed first-order kinetics. A substantial amount of SO remains even after 20 min of standing.

cannot form an olefin, and the energy required to form the $1°$ carbocation intermediate in the dehydration of ethanol is relatively large. Interferences from nitrite ion and C_3-to-C_5 alcohols are not expected to be significant in reversed-phase LC of sulfur compounds because they will elute in or near the void volume. Deliberately added NO_2^- has been considered as a void marker in LC analysis by this method.

Detection limits are compound dependent: sulfides, disulfides, and thiols are detected with nearly equimolar sensitivity on a per sulfur basis, but oxidized species show differing response factors. Detection limits, based on 95% confidence interval, are approximately 25 ppb S for reduced sulfur compounds in water. Bovine serum albumin, a ubiquitous protein used in cellular osmoregulation, has been detected [signal/rms (root mean square) noise = 15] at 200 nM concentrations in a 20 μL injection.

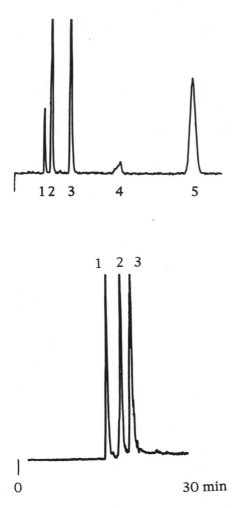

Figure 1.9. Ion-pair reversed-phase LC separation of a 4.0 nmol mixture of blood thiols detected by aqueous phase oxidation followed by CL detection: (1) N-acetyl-L-cysteine; (2) glutathione; (3) cysteine; (4) methionine; (5) dimeric glutathione. Separated on a 25 cm Rainin Microsorb MV C18 column packed with 5 μm particles; 1.0 mL/min flow rate of 2% CH_3CN/ 98% H_2O adjusted to pH = 2.6 with H_3PO_4, with 2 mM 1-octanesulfonic acid added as ion-pairing agent and 5 mM ethylenediaminetetraacetic acid (EDTA) added as preservative.

Figure 1.10. Reversed-phase HPLC separation of a protein mixture with LPSA detection: (1) 5.6 nmol insulin; (2) 2.7 nmol lysozyme; (3) 0.42 nmol bovine serum albumin. Injection, 10 μL; separation was accomplished on an Alltech HEMA-RP 150 × 4.6 mm C8 column packed with 10 μm particles. A linear mobile phase gradient of 90/10 A/B at $t = 0$ to 35/65 A/B at $t = 30$ min was used, where A was 0.15% trifluoroacetic acid in distilled water and B was 95% acetonitrile/5% distilled water. Total system backpressure was 200 atm; reactor temperature was 300 °C.

Aqueous phase conversion followed by $SO + O_3$ CL has been used to sensitively detect sulfur-containing pesticides, proteins, and amino acids, as well as inorganic sulfur compounds, at sub-ppm levels. Phenylthiohydantoin derivatives of both sulfur-containing and non-sulfur-containing amino acids can be sensitively detected owing to the presence of sulfur in the derivatizing agent. Application of this detector to LC analyses have included separations of blood thiols (see Figure 1.9), penicillamine from penicillamine disulfide, proteins (see Figure 1.10), and the separation of the sulfur-containing N-methylcarbamate pesticides listed in EPA (U.S. Environmental Protection Agency) Method 531.1. Acid-soluble thiols in rat plasma are also selectively

S1   S2

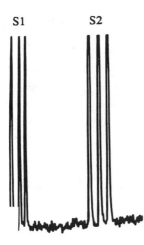

Figure 1.11. FIA of acid-soluble thiols in rat plasma using aqueous phase oxidation with CL detection. Blood samples (provided by Dr. B. Hybertson, Webb-Waring Lung Institute, Denver) were obtained from two anaesthetized male Sprague–Dowley rats (labeled S1 and S2), treated with perchloric acid to precipitate any proteins, centrifuged at 10,000 rpm for 15 min, the clear supernatant drawn off, and three 20 μL aliquots analyzed from each rat. Analysis conditions were as follows: 1.0 mL/min distilled water adjusted to pH = 2.31 with H_3PO_4 was used as the eluent, 300 °C reactor, 3000 psig backpressure, CL chamber maintained at 10 torr pressure. By comparison of the responses obtained from the plasma with responses from blood thiol standard solutions, the concentrations in the plasma are calculated to be 43 and 66 μM for the two rats S1 and S2, respectively, by the presently reported method. Relative standard deviations are less than 5% by this technique.

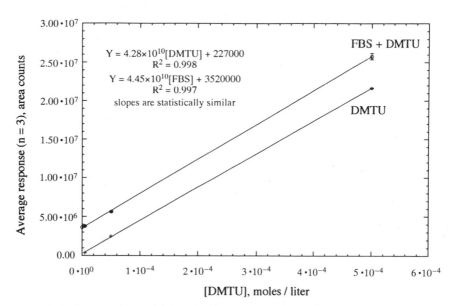

Figure 1.12. Aqueous phase oxidation of sulfur compounds is not affected by the complex matrix of bovine serum, as shown in this standard additions experiment. The upper curve shows responses for concentrations of DMTU spiked into bovine serum [$n = 3$ at each point; plotted are averages ±1 SD (standard deviation)]. The lower curve shows responses for identical concentrations of DMTU in distilled water. The curves are offset to the presence of sulfur-containing proteins in the serum. The slopes of the two curves are statistically similar, demonstrating the absence of matrix effects on detector response due to the serum. (DMTU = dimethyl thiourea; FBS = fetal bovine serum with 50 μM DMTU added.)

Table 1.2. Characteristics of Various Methods for Generating and Detecting Sulfur Monoxide

	Direct Ozone Oxidation	Flame SCD	Flameless SCD	SNPD	Aqueous Phase Oxidation
Detection limit, pg S/s	$\sim 1000^a$	0.8^b	0.05^c	2^d	16^e
Linearity, decades	Poor	5	3	3	3
Interferences	alkenes, PAHs, several others	—^{f}	—^{f}	Coeluting compounds at > 10 wt%	NO_2^-, C_3-to-C_5 alcohols
Selectivity vs. hydrocarbons	$> 10^7$	$> 10^7$	$> 10^8$	$> 10^6$	$> 10^{6g}$
Equimolar response	No	Yes	Yes	No	No

^{a}As thiophene.[62]
^{b}As thiophene.[49]
^{c}As 3-methylthiophene.[13]
^{d}As 2,5-dimethylthiophene.[14]
^{e}As dimethylthiourea.[15]
^{f}None reported.
^{g}Selectivity determined vs. nitrate ion and urea.[15]

detected in flow injection analysis (FIA) by this instrument (see Figure 1.11). Selectivity of the LPSA is sufficient that sulfur analysis in complex biological matrices exhibit no detectable matrix effects, as shown by the nearly identical slopes in the standard additions curves in Figure 1.12.

Flame conversion, adequate for microliter/minute flows in capillary LC, cannot easily be interfaced with analytical or preparative LC. Liquid phase oxidation of sulfur compounds permits much higher liquid flow rates, up to 10 mL/min, to be used in conjunction with $SO + O_3$CL detection in liquid chromatography.

A brief comparison of detection limits and characteristics of the five $SO + O_3$-based detection schemes is presented in Table 1.2.

REFERENCES

1. R. S. Hutte and J. D. Ray, in *Detectors for Capillary Chromatography* (H. H. Hill and D. G. McMinn eds.), Vol. 121, pp. 193–218. Wiley, New York, 1992.

2. S. A. Nyarady, R. M. Barkley, and R. E. Sievers, *Anal. Chem.* **57**, 2074–2079 (1985).

3. W. Breuning and F. J. M. Concha, *J. Chromatogr.* **122**, 253–265 (1975).

4. S. R. Spurlin and E. S. Yeung, *Anal. Chem.* **57**, 1223–1227 (1985).

5. S. O. Farwell and C. J. Barinaga, *J. Chromatogr. Sci.* **24**, 483–494 (1986).

6. J. K. Nelson, R. L. Getty, and J. W. Birks, *Anal. Chem.* **55**, 1767–1770 (1983).

7. E. A. Mishalanie and J. W. Birks, *Anal. Chem.* **58**, 918–923 (1986).

8. R. J. Glinski, E. A. Mishalanie, and J. W. Birks, *J. Photochem.* **37**, 217–231 (1987).

9. W. T. Foreman, C. L. Shellum, J. W. Birks, and R. E. Sievers, *J. Chromatogr.* **465**, 23–33 (1989).

10. C. J. Halstead and B. A. Thrush, *Photochem. Photobiol.* **4**, 1007–1013 (1965).

11. J. W. Birks and M. C. Kuge, *Anal. Chem.* **52**, 897–901 (1980).

12. R. L. Benner and D. H. Stedman, *Anal. Chem.* **61**, 1268–1271 (1989); R. Godec, N. Johansen, and D. H. Stedman, U. S. Patent 5, 330, 714 (July 19, 1994).

13. R. L. Shearer, *Anal. Chem.* **64**, 2192–2196 (1992).

14. T. B. Ryerson, R. M. Barkley, and R. E. Sievers, *J. Chromatogr. A.*, **670**, 117–126 (1994).

15. T. B. Ryerson, A. J. Dunham, R. M. Barkley, and R. E. Sievers, *Anal. Chem.* **66**, 2841–2851 (1994).

16. A. A. Turnipseed and J. W. Birks, in *Chemiluminescence and Photochemical Reaction Detection in Chromatography* (J. W. Birks, ed.), pp 39–69. VCH Publishers, New York, 1989.

17. W. L. Crider, *Anal. Chem.* **37**, 1770–1773 (1965).

18. R M. Dagnall, K. C. Thompson, and T. S. West, *Analyst* **92**, 506–512 (1967).

19. A. Syty and J. A. Dean, *Appl. Opt.* **7**, 1331–1336 (1968).

20. S. S. Brody and J. E. Chaney, *J. Gas Chromatogr.* **4**, 44–46 (1966).

21. C. H. I. Muller, K. Schofield, M. Steinberg, and H. P. Broida, *Symp. (Inter.) Combust., [Proc.]*, **17** (1979).

22. P. T. Gilbert, in *Analytical Flame Spectroscopy* (R. Mavrodineau, ed.), pp. 278–296. Springer-Verlag, New York, 1970.

23. C. F. Cullis and M. F. R. Mulcahy, *Combust. Flame* **18**, 225–292 (1972).

24. P. L. Patterson, R. L. Howe, and A. Abu-Shumays, *Anal. Chem.* **50**, 339–344 (1978).

25. T. Sugiyama, Y. Suzuki, and T. Takeuchi, *J. Chromatogr.* **77**, 309–316 (1973).

26. T. Sugiyama, Y. Suzuki, and T. Takeuchi, *J. Chromatogr.* **85**, 45–51 (1973).

27. P. L. Patterson, *Anal. Chem.* **50**, 345–348 (1978).

28. K. K. Gaines, W. H. Chatham, and S. O. Farwell, *J. High Res. Chromatogr.* **13**, 489–493 (1990).

29. J. F. McGaughey and S. K. Gangwal, *Anal. Chem.* **52**, 2079–2083 (1980).

30. T. Sugiyama, Y. Suzuki, and T. Takeuchi, *J. Chromatogr.* **80**, 61–67 (1973).

31. A. R. Baig, C. J. Cowper, and P. A. Gibbons, *Chromatographia* **16**, 297–300 (1982).

32. W. L. Crider, N. P. Barkley, M. J. Knott, and R. W. J. Slater, *Anal. Chim. Acta* **47**, 237–241 (1969).

33. M. Dressler, *Selective Gas Chromatographic Detectors*, J. Chromatogr. Libr., Vol. 36, Elsevier, New York, 1986.

34. B. G. Julin, H. W. Vandenborn, and J. J. Kirkland, *J. Chromatogr.* **112**, 443–453 (1975).

35. C. E. Kientz, A. Verweij, H. L. Boter, A. Poppema, R. W. Frei, G. J. De Jong, and U. A. T. Brinkman, *J. Chromatogr.* **467**, 385–394 (1989).

36. W. Frenzel, D. Schepers, and G. Schulze, *Anal. Chim. Acta* **277**, 103–111 (1993).

37. L. A. Pekay, and S. V. Olesik, *Anal. Chem.* **61**, 2616–2624 (1989).

38. H. Koizumi and Y. Suzuki, *J. High Resolut. Chromatogr.* **14**, 257–261 (1991).

39. R. F. Zainullin and V. G. Berezkin, *CRC Crit. Rev. Anal. Chem.* **22**, 183–199 (1991).

40. R. J. Glinski, J. N. Getty, and J. W. Birks, *Chem. Phys. Lett.* **117**, 359–364 (1985).

41. T. G. Chasteen, R. Fall, J. W. Birks, H. R. Martin, and R. J. Glinski, *Chromatographia* **31**, 342–346 (1991).

42. T. G. Chasteen, G. M. Silver, J. W. Birks, and R. Fall, *Chromatographia* **30**, 181–185 (1990).

43. J. K. Nelson, Fluorine-induced chemiluminescence detection of reduced sulfur compounds in gas chromatography. Doctoral Dissertation, University of Colorado, Boulder (1984).

44. M. A. A. Clyne, C. J. Halstead, and B. A. Thrush, *Proc. R. Soc. London, Ser. A* **295**, 355–362 (1966).

45. C. J. Halstead and B. A. Thrush, *Proc. R. Soc. London, Ser. A* **295**, 380–398 (1966).

46. C. J. Halstead and B. A. Thrush, *Proc. R. Soc. London, Ser. A* **295**, 363–379 (1966).

47. SCD is a registered trademark issued to Sievers Instrument Co., 2500 Central Ave., Boulder, Co.

48. P. W. Schenk and R. Steudel, *Angew. Chem., Int. Ed. Engl.* **4**, 402–409 (1965).

49. R. L. Shearer, D. L. O'Neal, R. Rios, and M. D. Baker, *J. Chromatogr. Sci.* **28**, 24–28 (1990).

50. H. Akimoto, B. J. Finlayson, and J. N. Pitts, Jr., *Chem. Phys. Lett.* **12**, 199–202 (1971).

51. R. J. Glinski, J. A. Sedarski, and D. A. Dixon, *J. Phys. Chem.* **85**, 2440–2443 (1981).

52. R. J. Glinski, J. A. Sedarski, and D. A. Dixon, *J. Am. Chem. Soc.* **104**, 1126–1128 (1982).

53. P. K. Arora and J. P. S. Chatha, *Can. J. Chem.* **62**, 417–423 (1984).

54. R. J. Glinski and D. A. Dixon, *J. Phys. Chem.* **89**, 33–38 (1985).

55. K. H. Becker, M. A. Inocêncio, and U. Schurath, in *Chemical Kinetics Data for the Upper and Lower Atmosphere*. Wiley, Warrenton, VA, 1974.

56. T. J. Kelly, J. S. Gaffney, M. F. Phillips, and R. L. Tanner, *Anal. Chem.* **55**, 135–138 (1983).

57. R. I. Martinez and J. T. Herron, *Int. J. Chem. Kinet.* **10**, 433–452 (1978).

58. J. S. Gaffney, D. J. Spandau, T. J. Kelly, and R. L. Tanner, *J. Chromatogr.* **347**, 121–127 (1985).

59. W. Breuning and F. J. M. Concha, *J. Chromatogr.* **142**, 191–201 (1977).

60. M. R. Zachariah and O. I. Smith, *Combust. Flame* **69**, 125–139 (1987).

61. R. Godec, N. Johansen, and D. H. Stedman, U.S. Pat. 5,227,135 (1993).

62. R. L. Benner, Development of the sulfur chemiluminescence detector. Doctoral Dissertation, University of Denver, Denver, CO (1991).

63. G. Black, R. L. Sharpless, and T. G. Slanger, *Chem. Phys. Lett.* **90**, 55–58 (1982).

64. R. M. Fristrom and A. A. Westenberg, *Flame Structure.* McGraw-Hill, New York, 1965.

65. N. G. Johansen, R. S. Hutte, and M. F. Legier, in *Monitoring Water in the 1990's: Meeting New Challenges*, J. R. Hall and G. D. Glysson, eds, ASTM Vol. No. STP 1102. ASTM, Philadelphia, 1991.

66. N. G. Johansen and J. W. Birks, *Am. Lab. (Fairfield, Conn.)* **23**, 112–119 (1991).

67. B. Chawla and F. Di Sanzo, *J. Chromatogr.* **589**, 271–279 (1992).

68. R. S. Hutte, N. G. Johansen, and M. F. Legier, *J. High Resolut. Chromatogr.* **13**, 421–426 (1990).

69. M. S. Burmeister, C. J. Drummond, E. A. Pfisterer, D. W. Hysert, Y. O. Sin, K. J. Sime, and D. B. Hawthorne, *J. Am. Soc. Brew. Chem.* **50**, 53–58 (1992).

70. H.-C. K. Chang and L. T. Taylor, *J. Chromatogr.* **517**, 491–501 (1990).

71. H.-C. K. Chang and L. T. Taylor, *Anal. Chem.* **63**, 486–490 (1991).

72. A. L. Howard and L. T. Taylor, *J. High Resolut. Chromatogr.* **14**, 785–794 (1991).

73. R. L. Benner and D. H. Stedman, *Environ. Sci. Technol.* **24**, 1592–1596 (1990).

74. A. L. Howard and L. T. Taylor, *Anal. Chem.* **65**, 724–729 (1993).

75. S. E. Eckert-Tilotta, S. B. Hawthorne, and D. J. Miller, *J. Chromatogr.* **591**, 313–323 (1992).

76. R. L. Shearer, E. B. Poole, and J. B. Nowalk, *J. Chromatogr. Sci.* **31**, 82–87 (1993).

77. D. D. Bombick and J. Allison, *J. Chromatogr. Sci.* **27**, 612–619 (1989).

78. L. Trelli and D. Bombick, *Pittsburgh Conf. Anal. Spectrosc.* New Orleans, LA, Poster No. 148P (1992).

79. G. E. Hartzell and J. N. Paige, *J. Am. Chem. Soc.* **88**, 2616–2617 (1966).

80. A. J. Dunham, R. M. Barkely, and R. E. Sievers, *Anal. Chem.* in press.

81. S. Ramayya, S. Brittain, C. DeAlmeida, W. Mik, and J. J. Antal, *Fuel* **66**, 1364–1371 (1987).

CHAPTER

2

FLAMELESS SULFUR CHEMILUMINESCENCE DETECTION

RANDALL LEE SHEARER

*Shell Development Company,
Houston, Texas 77083**

2.1. INTRODUCTION

In 1983 Drushel of Exxon described the needs of the chromatographer in the area of detectors.[1] Drushel noted that because the quantitative aspects of chromatographic analyses are dependent upon detector capabilities, one of the most important needs of the chromatographer is to have good detectors. This is most assuredly as true today, despite the improvements made in the separation processes, which provide more resolving power than ever. Nevertheless, it is often sufficient to measure compound classes as opposed to exhaustive analysis of individual compounds in a sample. Selective detectors give one the ability to make such measurements. As Drushel pointed out, selective detectors are particularly useful in the petroleum industry, where in most cases exhaustive analysis is neither practical nor possible. Of course this is true in many areas of chemistry. Selective detectors make important contributions in the analysis of agricultural samples, biological tissues, foods, beverages, environmental samples, perfumes and fragrances, specialty chemical products, etc. Many of the changes occurring in the petrochemical industry today are being faced in other industries as well. This chapter is decidedly slanted toward the petroleum industry based on the experiences of the author, but the general content is relevant to other areas of chemistry. The aim here will be to review the development of the flameless sulfur chemiluminescence as a detection method for sulfur compound measurement by chromatography.

*Current address: Sievers Instruments 2500 Central Ave., Boulder, Colorado 80301.

Selective Detectors, edited by Robert E. Sievers. Chemical Analysis Series, Vol. 131.
ISBN 0-471-01343-9 © 1995 John Wiley & Sons, Inc.

In addition, it is hoped that this chapter will stimulate interest in deserving areas of study.

2.1.1. Need for Sulfur Compound Analysis in the Petroleum Industry

It has been known that sulfur compounds in crude oil are the single largest cause of corrosion in crude distillation units for over a hundred years. One of the oldest analytical tests for the presence of sulfur in petroleum products is based on coloring of a copper strip. This test is still used in refineries today.[2] Interestingly, the English word *sulfur* originated from the Sanskrit word *shulbari*, meaning enemy (*ari*) of copper (*shulba*).[3]

Today industry spends enormous capital and effort to treat the effects of sulfur corrosion and for sulfur control, abatement, and recovery. The expense and effort required for this will continue to rise. Refineries are, in general, being required to handle crude slates containing higher levels of sulfur. While there is currently an abundance of oil in the global market, there is a shortage of low-sulfur crude. Many refineries must undergo significant upgrades and metallurgical changes so that they can accept this increasingly poorer diet.

New governmental regulations are also challenging the petroleum and chemical industries to reduce emissions and to boost safety of consumer products. In order to better compete in the global marketplace, greater efforts are being made to improve consistency, reliability, and performance of all products. All of these demands are extremely challenging to the chemists and engineers that make the industry work.

From the pollution standpoint, the form of the sulfur in fuels is largely unimportant, since the sulfur compounds are combusted to sulfur dioxide. However, the presence of ppb levels of certain sulfur compounds may adversely impact the quality of many consumer products. Catalysts used in the industry are also frequently sensitive to ppm or even ppb levels of certain sulfur compounds. It is incorrect to generalize that sulfur compounds are always bad. Sometimes sulfur compounds may intentionally be added to some products in order to impart a specific property, e.g., use of odorants for natural gas and liquefied petroleum gases (LPGs). And of course many sulfur compounds themselves are valuable products. But it is safe to generalize that sulfur analysis traditionally has been difficult.

Instrumental methods for total sulfur analyses are very important and are widely used, but it should be recognized that the amounts present of specific sulfur compounds are frequently more important than sulfur content alone. For this reason, analytical techniques that can measure specific sulfur compounds are more informative than techniques that measure only sulfur content. This is why chromatographic techniques coupled with sulfur-selective detection are so useful. In addition, chromatographic measurements can often be made more easily. For example, the boiling distribution of sulfur compounds in numerous streams and products has been frequently determined by physical distillation followed by sulfur or elemental analysis of each cut. The development of simulated distillation with element specific detection is carried out more safely and easily and is less likely to introduce errors or contamination from sample handling.

2.1.2. Sulfur-Selective Detectors

A number of commercially available selective detectors have been used for the determination of sulfur compounds. These include the flame photometric detector (FPD),[1,4-6] electrolytic conductivity (Hall) detector,[1,6,7] atomic emission detector (AED),[6,8,9] and various chemiluminescence detectors.[6,10-15] The FPD is the most commonly used detector that is selective for sulfur compounds, but it suffers from numerous problems. The FPD is not well suited for quantitative analysis of sulfur compounds in complex hydrocarbon matrices because of hydrocarbon quenching and interferences. The FPD exhibits nonlinear (exponential) response to sulfur compounds, compound dependent response factors, and other "enigmas."[5] The FPD is, however, relatively inexpensive, relatively robust, and adequate for many applications.

The electrolytic conductivity detector is sometimes used for sulfur analysis, but in many instances it suffers from interferences similar to those of the FPD. In addition, the detector requires a significant amount of maintenance. It is relatively inexpensive and produces a linear response, however, and generally lower detection limits than those reported for the FPD.[6]

The AED and the flame sulfur chemiluminescence detector (SCD) have recently become commercially available and their performances have been compared.[16] In general, the AED and SCD perform similarly with regard to their figures of merit, but the flame SCD exhibits much better selectivity. The AED is better than the FPD for quantitative analysis of sulfur compounds in complex matrices. The AED does not exhibit as many problems with interferences and quenching, and sulfur response is compound independent. The greatest advantage of the AED is that multiple elements can be determined simultaneously. Trace level sulfur analysis is difficult using the AED, however, unless chromatographic resolution of trace components of interest from major components is achieved. The AED is also expensive, and its use requires skilled analysts. Nevertheless, the AED is one of the most powerful new analytical tools for selective detection in GC.

Selective detectors based on chemiluminescence are attractive because they are usually sensitive and inherently selective, since few compounds undergo chemiluminescent reactions. The flame SCD is finding widespread use for the analysis of trace levels of sulfur compounds in hydrocarbon matrices.[17-19]

2.2. PRINCIPLE OF OPERATION OF THE SULFUR CHEMILUMINESCENCE DETECTOR

Detection of sulfur compounds in the SCD is based upon a two-step mechanism as shown below. In the first step, all sulfur compounds entering the flame are combusted to form SO and SO_2

$$S\text{-compound} + H_2 + O_2 \rightarrow SO + \text{other products}$$

$$SO + O_3 \rightarrow SO_2 + O_2 + hv$$

(SO_2 is not detected). In the second step of the mechanism, light energy (hv) in the blue region of the spectrum is emitted from the chemiluminescent reaction of SO with ozone. In the conventional version of the instrument, the SO and other products are formed in a flame ionization detector (FID). The combustion products are sampled by means of a ceramic probe and transferred to a chemiluminescent detector, where ozone is added and the emitted light is measured by a photomultiplier tube (PMT).

The SCD possesses numerous advantages over other sulfur detectors for the measurement of low levels of sulfur compounds. The SCD is more sensitive than the FPD by at least a factor of 10. Sulfur compounds produce a linear response in the SCD that spans some 3–4 orders of magnitude. Since all sulfur compounds are converted equally to SO, they produce the same detector response. The presence of hydrocarbons coeluting with sulfur compounds causes little if any quenching in the SCD. The SCD possesses the best sulfur-to-carbon selectivity of any available sulfur selective detector ($> 10^6$).

Recently, a new way to generate SO for the SCD was developed,[20] and its application to detection of sulfur compounds in petroleum products has been described.[21] This new development is called flameless SCD, since an externally heated combustion assembly, instead of a flame, is used to generate SO.

2.2.1. Combustion Chemistry of Sulfur

Examining the combustion chemistry of sulfur is worthwhile in an effort to elucidate important processes involved in generating a chemiluminescent species in the SCD. When sulfur species are added to hydrogen-air flames, sulfur is distributed among the following species: SO_3, SO_2, SO, S, S_2, HS, and H_2S. Kramlich has shown that the relative ratios of SO_2, SO, H_2S, SH, and S are approximately $100:20:1:1:1$.[22] The concentrations of these species deviate significantly from equilibrium values depending on stoichiometry of the flames and where the flame gases are sampled.[23,24] In the flame reaction zone, SO_2 and SO predominate. Even in a 20% deficiency of air, 90% of the sulfur is present as SO_2 and the remainder exists almost entirely as SO. To explain this, it has been suggested that the concentration of the radicals H and OH are above equilibrium values in the reaction zone and that the following bimolecular reactions suppress the formation of H_2S, S_2, and SH[25,26]:

$$H_2 + SO_2 \rightleftharpoons H_2O + SO$$

$$H_2 + SO \rightleftharpoons H_2O + S$$

$$H_2 + S \rightleftharpoons SH + H$$

$$H_2 + SH \rightleftharpoons H_2S + H$$

$$SH + SO \rightleftharpoons S_2 + OH$$

In fact, [H] is several orders of magnitude greater than the equilibrium value of [H] in the reaction zone of a typical fuel-rich flame in the absence of sulfur.[27] Thus, at low sulfur concentrations such as those found in the flame SCD, the absence of H_2S, S_2, and SH in the reaction zone should be expected. Indeed, in hydrogen-rich flames, added sulfur is found to be mostly SO_2, as predicted by the following equilibrium reaction because of dissociation of molecular hydrogen to hydrogen atoms.[25]

$$ZH + H_2S + 2H_2O \rightleftharpoons SO_2 + 4H_2$$

It should be understood that, in general, radical intermediates such as SO, SH, H, and OH are frequently present in superequilibrium concentrations during combustion and that these species become more stable relative to conventional product molecules at high temperatures. As combustion gases cool, reaction rates decrease precipitously so that chemical species can be "frozen" if the temperature drop is sufficiently rapid. A similar effect is obtained upon expansion of combustion gases from a high-pressure to a low-pressure zone. In the flame SCD this is exactly what happens. The reaction zone gases are sampled by a probe that rapidly lowers the temperature of the exhaust gas to about room temperature and rapidly reduces its pressure to about 10 torr. This effectively freezes the composition of the flame gases where they are sampled. Flame residence time may be adjusted by changing the distance of the sampling probe relative to the hydrogen and air inlets and by adjusting hydrogen and air flow rates. Thus, different parts of the flame can be sampled and sensitivity can thus be optimized. In the flame SCD detection system, the flame residence time could be varied from 1 to 40 ms.[14]

Of the possible sulfur species formed in a flame, SO, H_2S, S, and SH all can react with ozone and produce chemiluminescence. Benner unequivocally showed that SO is the chemiluminescent precursor formed in the flame of the SCD.[28] A combination of spectral and kinetic measurements were made as follows. The chemiluminescence spectrum in the SCD was measured using a monochromator and GaAs PMT connected to the SCD reaction cell. The measurement of the emission spectrum from the flame source was characteristic of the $^1B-^1A_1$ transition of SO_2 described previously by other researchers.[29] The reaction of SH with ozone produces HSO* with an emission centered around 640 nm.[30] No emission was observed in this region, nor at the wavelengths (384 and 394 nm)[5] where the emission from excited S_2 would be expected. Independent spectroscopic analysis of the chemiluminescence occurring in the flame SCD also showed that excited SO_2 is the principal emitter.[31] These observations rule out the possibility that HS or S are present in analytically significant quantities in the SCD.

Benner showed that H_2S is not present in analytically significant quantities by making kinetic rate constant measurements of the chemiluminescent precursor formed in the flame of the SCD; SO and H_2S react with ozone to produce SO_2 in the same excited state.[32] The gas phase reaction of SO with ozone is very rapid,[33,34] however, whereas the reaction of H_2S is much slower.[35,36] The reported rate constants of H_2S with ozone are less than reported rate constants for SO with ozone by at least 2 orders of magnitude. Benner used a microwave discharge to generate SO and compared its measured reaction rate constant with ozone to that of the species generated by the flame of the SCD. The measured rate constants were equivalent within experimental error. This verifies that SO, and not H_2S, is the species responsible for the chemiluminescence in the SCD. Furthermore, Benner found that the chemiluminescent intensity in the flame SCD is reduced by addition of NO_2. In fact, this reagent was used to titrate SO in the gas phase to determine the efficiency of the SCD flame. It was found that approximately 4 out of every 1000 SO_2 molecules entering the flame reaches the chemiluminescence detector as SO. It should be pointed out that H_2S does not react with NO_2, so this is further confirmation that H_2S is not analytically important in the flame SCD.

2.2.2. Chemiluminescence Detection

While various chromatography detectors based on nonflame chemiluminescence have been used in academia since at least the early 1970s, only in the past 10 years or so have they become commercially available. This probably has something to do with the complexity of chemiluminescence detectors (need for an ozone generator, a vacuum pump, etc.) compared to commonly used detectors. On the other hand, chemiluminescence-based detectors are less complex than some commonly used selective detectors like the electrolytic conductivity (Hall) detector. In addition, chemiluminescence detectors like the nitrogen-selective detector, based on the chemiluminescence of nitric oxide with ozone, and the sulfur chemiluminescence detector demonstrate numerous advantages including unprecedented selectivity.[6,15]

Successful detection of analytes by chemiluminescence requires that the analytes chemiluminesce directly with a reagent gas or are transformed, by means of a reactor, etc., into a species that will do so. Carrier gases and matrix components should not interfere with the analysis. In fact, reactors prove useful in converting compounds that might interfere into ones that pass through the detector undetected. A number of other common factors affect the performance of detectors based on chemiluminescence. In the final result, the amount of light that is detected constitutes the basis of the analytical measurement. This measurement is dependent upon the actual photon flux generated, sensitivity of the light-measuring device used, optical collection efficiency, analyte and reagent mixing ratios, chemical yield of any reactor being used in the apparatus, and, of course, chromatographic efficiency (injection volumes, flow rates, etc). All of these factors must be optimized in concert in order to produce the best results. For example, large cell volumes favor better detection by increasing the optical collection efficiency. On the other hand, large volumes contribute to band broadening and loss of efficiency in chromatographic systems. Fortunately, low pressure enhances the light that can be detected by reducing losses from collisional quenching and at the same time reducing the dead volume in the chromatographic system. In general, the lower the pressure and the higher the pumping speed of the vacuum pump, the better the results.

Generation of light is pseudo-first-order with respect to the chemiluminescent species being measured because the reagent gas is used in large excess. Unless the chemiluminescent species in the cell is totally consumed by the reagent, then higher mixing ratios of the reagent can generate more light. Thus, the use of a more efficient ozone generator, for example, may lead to enhanced sensitivity. Some real-time instruments based on chemiluminescence utilize very large reaction cells and large throughput volumes. This yields extremely sensitive measurements. When used as a chromatographic detector, however, the system is constrained by typical chromatographic flow rates and the need to avoid extracolumn band broadening. Our understanding of the basics of these detection systems has advanced considerably in recent years. A thorough discussion of the fundamentals and potential of chemiluminescence detectors applied to chromatography is beyond the scope of this review, but the topic is addressed in detail elsewhere.[15,37]

2.3. FLAMELESS SULFUR CHEMILUMINESCENCE DETECTION

The flame SCD is nearly an ideal sulfur-selective detector, and its use is especially advantageous for detection of sulfur compounds at low levels in complex matrices. The flame SCD produces a nearly equimolar response to sulfur compounds. It does not exhibit significant quenching of response from other compounds, it possesses high sensitivity and selectivity, and a linear response to sulfur over 3–4 orders of magnitude is obtained. Nevertheless, the flame SCD, like most selective detectors, possesses some limitations that hamper its use.

The flame SCD detector response is dependent on the condition and positioning of a ceramic probe that is used to sample the postflame gases from a flame ionization detector. Commonly, the analyst will have to carefully adjust probe position and gas flow rates in order to achieve optimum detector performance. However, detector response may change, or the probe may even become plugged, when column bleed or other impurities accumulate on the probe, and this requires the analyst to condition or change the probe in order to maintain optimum detector performance. Conditioning of the probe may become time consuming, but generally users of the detector become more proficient in this operation and learn techniques to make the task simpler and to avoid these problems in the first place. For example, high-bleed columns and high column temperatures should be avoided. Carrier and detector gases should be of good quality and should be further purified using molecular sieve or other appropriate traps. Thick-film capillary columns, while providing good performance for light sulfur compounds, also tend to exhibit considerable bleeding at temperatures greater than 250 °C. Column bleed causes problem with other detectors [FPD, ELCD (electrolytic conductivity detector), ECD (electron capture detector), and MS (mass spectrometer), in particular] and column manufacturers have recently begun to produce low-bleed thick-film phases. It is important that oxygen be removed from the carrier gas and that new columns be conditioned before being used with the flame SCD. In no case should column-rejuvenating solutions or silanizing agents be injected into a column connected to an SCD.

It is believed that the detector response is altered through the buildup of materials (possibly silicon dioxide or silicates) that either prevent the formation of SO or catalyze its destruction. Although these problems can be corrected and are largely avoidable, the exact cause of them is not yet clear. This characteristic has limited flame SCD use to skilled operators and has prevented it from being used in process instruments. The excellent results achieved with the flame SCD and the desire to improve its stability have prompted an examination of other ways to generate SO.

About a dozen methods are reported in the literature describing the formation of SO from other sulfur compounds.[38] For example, SO is formed from other sulfur oxides by ultraviolet (UV) and gamma ray photolysis, microwave discharge, glow discharge, and thermal dissociation. It is formed from sulfides and sulfides by roasting and thermolysis. Photolysis of hydrogen sulfide with molecular oxygen and reaction of hydrogen sulfide with atomic oxygen have been used to form SO, and, from elemental sulfur, it is formed by combustion with oxygen, combustion with CuO, heating with SO_2, and electrical discharge in oxygen.

2.3.1. Differences from Flame-Based Sulfur Chemiluminescence Detection

The flameless SCD uses an externally heated ceramic combustion assembly, operated at low pressure and under fuel-rich conditions, to generate SO. A drawing of the burner (as manufactured by Sievers Instruments, Inc., Boulder, Colorado) is shown in Figure 2.1. The burner essentially consists of a ceramic tube (1.32 mm o.d.; 0.51 mm i.d.) within a ceramic tube (3.17 mm o.d.; 1.6 mm i.d.). The inner tube extends approximately 3 cm into the top of the outer tube. A gas chromatography (GC) column or supercritical fluid chromatography (SFC) restrictor is inserted approximately 1 cm into the outer tube from the bottom. This design is aimed at increasing detector reliability and sensitivity. In the burner, hydrogen and air are mixed as the column effluent reaches the hottest zone of the reactor. Thus, in a single furnace, partial oxidation of the effluent from the column is possible before the stream reaches the hottest zone in the reactor. This helps prevent formation of coke in the reactor. Significantly more coke is formed from hydrocarbons if the air and hydrogen flow paths are reversed.

Problems related to deposition of impurities within the probe are greatly reduced, and plugging of the flow paths in the furnace is not possible (at least none has occurred in more than 2 years of operation of multiple instruments). The combustion products are swept to the chemiluminescence analyzer through the annular space between the inner and outer ceramic tubes of the burner assembly. This provides much greater surface area than in the flame SCD, such that deposition of contaminants should have less effect. This design allows for the deposition of contaminants at the base of the outer tube before they reach the hydrogen–oxygen combustion zone, where the active chemiluminescent species is formed. The following reaction is thought to be important in the production of sulfur monoxide in both SCD

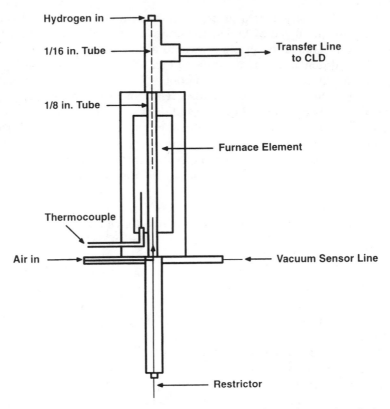

Figure 2.1. Flameless SCD burner. A restrictor (or column) is inserted approximately 1 cm into the outer ceramic tube from the base of the burner. Hydrogen enters the top of the burner through a $\frac{1}{16}$ in. o.d. ceramic tube. Combustion products are swept to the chemiluminescence detector (CLD) in the annular space between the outer and inner tubes.

designs.[14] Because the flameless system

$$SO_2 + H_2 \rightleftarrows SO + H_2O$$

is enclosed and external heating is used to maintain elevated temperatures, hydrogen/air ratios outside of the normal flammability limits can be utilized. At the temperatures employed in the flameless SCD, hydrogen and oxygen may behave as active species on the surface of the ceramic tubes.

Because the tubes in the flameless SCD are enclosed, they are less susceptible to physical breakage. Some deterioration of the inner tube is observed when the reactor is operated at 1100–1200 °C, but little occurs when the reactor is operated at temperatures of 1000 °C or less. This flameless SCD system utilizes combustion at low pressure, nominally 10–200 torr, and high temperature, nominally 750–900 °C. It is possible that this low-pressure combustion results in increased production and preservation of sulfur monoxide, since any SO formed is less likely to be lost to subsequent reactions. The design leads to lower pressures in the chemiluminescent reaction cell, and this alone should lead to improved sensitivity, since the intensity of the chemiluminescence is known to vary inversely with pressure.[30] The total system is enclosed, and therefore all of the combustion products are collected and transferred to the chemiluminescent reaction cell. This is not necessarily the case in the flame SCD, in which hydrogen and air flow rates must be carefully optimized to achieve a collection efficiency near unity. In fact, in the flame SCD, if the sum of the hydrogen and air flow rates is too low, then ambient air will be drawn into the chemiluminescence analyzer. This problem cannot occur in the flameless SCD, regardless of hydrogen or air flow rates, unless the system leaks. Atmospheric pressure changes can affect the collection efficiency of the probe in the flame SCD; however, the flameless system is insensitive to atmospheric pressure swings. This would be particularly important for instruments designed to make measurements from aircraft. The fact that the detector is enclosed is also desirable for on-line process instrumentation.

2.3.2. Operation and Performance

Flameless SCD is compatible with both packed and capillary columns. Figure 2.2 shows a block diagram of the flameless SCD coupled with a gas chromatograph. The detector uses hydrogen and air for the burner and oxygen for the ozone generator (air may be used, but sensitivity is somewhat lower, as will be discussed later). The interface controller regulates gas flow rates to the burner as well as its temperature. A pressure transducer is used to monitor pressure in the burner. As a safety feature, the controller shuts off hydrogen flow and electric current to the burner if a leak is detected. A Teflon transfer line under reduced pressure is used to carry the combustion products to a chemiluminescence detector. The chemiluminescence detector consists of a reaction cell, an optical filter, an ozone generator, a PMT, chemical traps, a vacuum pump, and associated electronics. Table 2.1 gives typical operating conditions of the detector.

Considering the typical conditions encountered in the flameless SCD, its excellent performance is surprising in light of the very short time the column effluent spends in the burner. In fact, the reaction zone probably has a much smaller volume and the residence time of an analyte in the reaction zone is probably ≪ 1 ms. Of course this short residence time is important for generation and preservation of SO and for maintaining chromatographic peak integrity. As mentioned previously, Benner used gas phase titration of

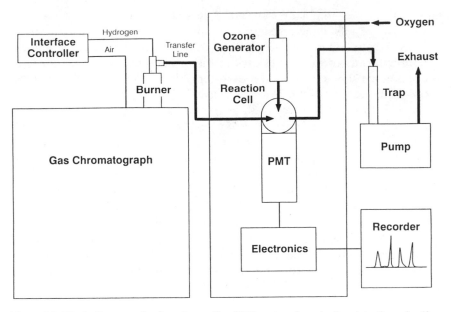

Figure 2.2. Block diagram of a flameless sulfur CLD system for selective detection of sulfur-containing compounds by gas chromatography.

Table 2.1. Typical Flameless SCD Operating Conditions

Hydrogen flow rate to burner	100 mL/min
Air flow rate to burner	25 mL/min
Temperature of burner	775 °C
Internal volume of burner	0.12 cm^3
Pressure in burner	100 torr
Burner residence time	2 ms
Pressure in chemiluminescence cell	10 torr
Ozone ($\sim$ 5%) flow rate to chemiluminescence cell	50 mL/min

SO with NO_2 to determine that about 0.4% of sulfur molecules entering the flame in a quartz burner reaches the reaction cell as SO.[28] If we assume that 10–15% (the theoretical maximum possible concentration of SO in flames)[22,23] of the sulfur entering the detector can be converted to SO, then the potential for improved detection limits is readily apparent.

While the chemistry occurring in the flameless burner is undoubtedly very similar to that occurring in the flame that was well characterized by Benner and Stedman, MS was used to analyze the major products of the flameless burner and to gain further insight into the mechanisms of this detector. In the experiment, a 100 μm i.d. fused silica capillary tube was used to sample the gases generated in a flameless burner in parallel with a chemiluminescent detector. An HP-5971 mass selective detector and an HP-Chemstation (Hewlett Packard) was used for mass spectral analysis. As has been found previously with flame-based combustion, the major sulfur species produced in the flameless burner are SO and SO_2. The burner was operated under

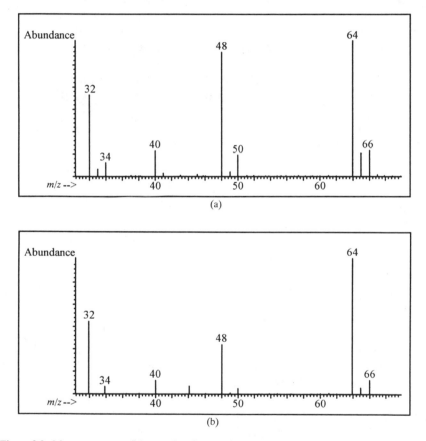

Figure 2.3. Mass spectrum of the combustion products generated from hydrogen sulfide being introduced into the flameless burner (a). The mass spectrum is identical to that of sulfur dioxide (b), with two exceptions. The ratio of m/z 48 to m/z 64 (here m = mass; z = charge) is higher in the spectra of the hydrogen sulfide combustion products than in that of sulfur dioxide measured directly. Additional abundance of m/z 48 is from the presence of sulfur monoxide in the combustion products. The m/z 40 is from argon present in the carrier gas.

typical conditions (105 mL/min hydrogen, 25 mL/min air, and 775 °C). The combustion products of methyl mercaptan, hydrogen sulfide, sulfur dioxide, and hydrocarbons were examined. Essentially the same mass spectrum was obtained for each sulfur compound except that CO_2 was found in the mass spectrum from methyl mercaptan, as would be expected. Figure 2.3 shows the mass spectrum of the combustion products generated from H_2S introduced into the flameless burner. The mass spectrum is identical to that of SO_2, with two exceptions. The ratio of m/z 48 to m/z 64 is higher in the spectrum of the H_2S combustion products than in that of SO_2 measured directly. The m/z 40 is from argon present in the carrier gas. The additional m/z 48 is from the presence of SO in the combustion products. As would be expected, only CO_2 was observed as the main carbon-containing combustion product of hydrocarbons, although the mass range below m/z 30 was not scanned.

It ought to be possible to determine the fraction of SO being formed were its ionization efficiency in the mass spectrometer known (supposing that a known SO generator were available in our laboratory), but this approach has not yet been pursued. Nevertheless, by assuming comparable ionization efficiencies for SO and SO_2, it is estimated that nominally 1–10% of the sulfur is in the form of SO. The lack of measurable m/z 35 and 36 demonstrates that H_2S is not present at detectable levels (m/z 34 is due to the ^{34}S isotope resulting from fragmentation of SO and SO_2). This also explains why the chemiluminescence signal obtained directly from H_2S is enhanced by a factor of about 30 after it is passed through a flame or burner.[20]

2.3.2.1. Sensitivity

As for the GC detector, minimum detectable levels on the order of 25–50 fg S/s are routinely achieved on some instruments when oxygen, rather than air, is used for the ozone generator.[20] A minimum detectable level as low as 10 fg S/s has been observed, but not repeatably.[20] The manufacturer of the flameless SCD sets a specification for minimum detectability at < 0.5 pg S/s. The manufacturer's specification for minimum detectability of the flame SCD is < 5 pg S/s. Thus, the flameless SCD demonstrates better sensitivity than the flame SCD by approximately an order of magnitude. The sensitivity of flameless SCD applied to SFC is essentially equivalent to that obtained in GC.[39]

Further study will be needed to determine whether this is a result of greater production of SO, from reduced collisional quenching of excited SO_2, or from some other effect. In the early development of flameless SCD, its absolute sensitivity appeared to be at a maximum at temperatures above 1000 °C.[20] Background noise was also found to increase with increasing burner temperature, so that the optimum signal to noise was observed at

800–900 °C. A practical operating temperature of 800 °C was chosen, since the burner is expected to have a greater lifetime at lower temperatures. Burner temperatures as low as 600–700 °C produce reasonable detector response to sulfur-containing compounds. Little response is observed below these temperatures, although some sulfur compounds will chemiluminescence directly with ozone.[13] It should be noted that temperatures were measured at the surface of the outer ceramic tube in this early work and small differences in the positioning of the measuring thermocouple could account for errors in temperature measurements of up to 50 °C. It should also be noted that the temperature in the reaction zone in the burner could be very much different (probably hotter) from the measured temperature at the outside of the tube. In any case, detector response exhibits a broad dependence on burner temperature, i.e., no sharp optimum. This is a desirable feature in that it is indicative of stability of the detector.

Similarly, the flameless SCD response exhibits a broad dependence on air and hydrogen flow rates to the burner. Optimum flow rates for hydrogen and air were determined to be approximately 80 mL/min for hydrogen and 20 mL/min for the original system.[20] The response curve was broad for both hydrogen and air, but the air flow rate had a much greater effect on detector response. When the hydrogen flow rate was held at 80 mL/min, a substantial decline in sensitivity was observed by raising the air flow rate from 50 to 60 mL/min. Nevertheless, as long as hydrogen and air flow rates remained stable, good detector performance was observed over a wide range of conditions.

No significant difference in detector performance was observed when oxygen was used instead of air in the flameless system, as long as the stoichiometry of hydrogen and oxygen was the same. It is easier, however, to accurately and precisely regulate larger air flow rates than smaller oxygen flow rates. Interestingly, in the flame SCD, pure oxygen cannot be used because the resulting flame is too hot and could cause materials used in the FID to ignite. A small amount of nitric oxide is continuously formed in air flames, and the background chemiluminescence from the reaction of nitric oxide with ozone is eliminated with a UV bandpass filter. A potential advantage of the flameless system using oxygen instead of air in the burner is that it can be operated without this UV bandpass filter, since nitric oxide will not be formed. In fact, a modest gain in S/N (signal-to-noise ratio), some 10–20%, is observed, although sulfur compound to nitrogen compound selectivity is diminished to about 10^4.

As an example of the level of detection that can be achieved in flameless SCD, Figure 2.4 shows a chromatogram from a 1 μL splitless injection of 1 ppb levels of sulfur compounds in n-hexane. Each peak at the 1 ppb level represents approximately 0.7 pg of sulfur injected. A minimum detectable

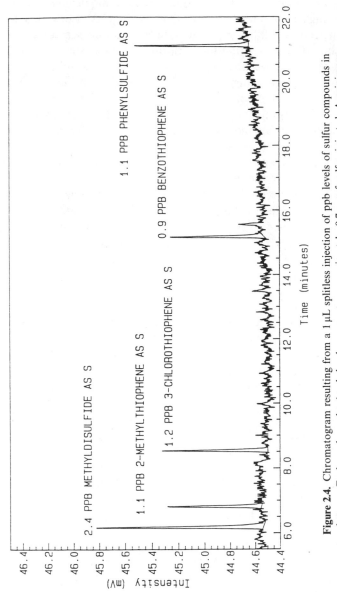

Figure 2.4. Chromatogram resulting from a 1 μL splitless injection of ppb levels of sulfur compounds in *n*-hexane. Each peak at the 1 ppb level represents approximately 0.7 pg of sulfur injected. A minimum detectable level of approximately 25 fg S/s at $S/N = 3$ is demonstrated. Reprinted from Shearer,[20] with permission.

level of approximately $25\,\text{fg}\,\text{S/s}$ at $S/N = 3$ is demonstrated. Thus, in its practical application, sub-ppm levels of sulfur compounds can be detected by using a conventional split/splitless injection port and normal injection sizes, etc. Selectivity of the detector, in essence, is enhanced because of the excellent sensitivity of the detector i.e., it is not necessary to employ pre-concentration techniques or to inject large volumes of sample matrices to detect sulfur compounds at low concentrations. Therefore, sample volumes are easily kept below levels that would cause a detector interference or false response.

It is particularly worth noting that the flameless SCD is more sensitive to sulfur than the FID is to carbon, approximately $1\,\text{pg}\,\text{C/s}$.[40] Therefore, even in cases in which sulfur analytes are well separated from matrix components, methods based on the flameless SCD possess the advantage of lower detection limits. The use of sulfur-containing derivatization reagents to make analytes amenable to detection by flameless SCD is an area that should be exploited. It should also be noted that the sensitivity of individual detectors may vary by a factor of 2 or more because of differences in burner efficiency, PMT gain and detector electronics, ozone generator efficiency, and the like. Column bleed is known to cause a reduction in detector sensitivity, but the mechanism of this decline is not understood and should be investigated further.

2.3.2.2 Selectivity

Hydrocarbons are not detected under normal conditions, e.g., typical injection sizes. Some response will be observed if the ability of the burner to completely combust the column effluent is exceeded. This might happen if the rate of introduction of hydrocarbon into the burner is greater than approximately $0.1\,\text{mg}\,\text{C/s}$. Typically, no response is observed from the injection of solvents or other hydrocarbon products that do not contain sulfur. Sulfur-to-carbon selectivity is typically greater than 10^8 on a weight/weight basis. As mentioned previously, nitrogen-containing compounds will produce a weak response in the flameless SCD because they are converted to nitric oxide in the burner and, thus, some chemiluminescence from the reaction of nitric oxide with ozone may be detected. The major chemiluminescence of that reaction is in the infrared region of the spectrum, however, so this interference is small. The sulfur-to-nitrogen selectivity is on the order of 10^5 on a mass basis. In typical petroleum related samples, nitrogen-containing compounds are typically present at lower levels than sulfur-containing compounds, so there are few interferences from nitrogen-containing compounds. In SFC, no response was obtained from solvents that were free of sulfur compounds and it was concluded that selectivity was at least 10^6 for sulfur over carbon.[39]

2.3.2.3. Relative Response Factors

If all sulfur compounds are converted into a common species that is detected by its chemiluminescence, then all sulfur compounds should exhibit the same response on a sulfur basis; this has been observed.[20,21] The advantage of compound-independent response is that calibration of response from a single component could be used to quantitate all sulfur compounds in a sample. It should be noted, however, that many sulfur compounds are reactive and easily lost due to oxidation to another sulfur species or to surface adsorption, especially on active sites in valves and inlets. Chromatographic problems such as sample split discrimination can also affect response factors. Therefore, careful attention must be paid to all sample handling, standard preparation, and chromatographic technique, in order to take advantage of the equimolar detector response. Care should also be taken to ensure that peaks of sulfur compounds are not off-scale and are within the linear range of the detector.

Response factors relative to thiophene, in arbitrary units of area response per unit mass of sulfur, for a number of model sulfur compounds are shown in Table 2.2. These compounds represent compounds classes (mercaptans, sulfides, disulfides, and thiophenes) that are typically encountered in petroleum-related samples. The response factors were obtained by on-column injection with a 0.53 mm i.d. SPB-5 column. The detector was operated according to the manufacturer's instruction (775 °C burner temperature; 100 mL/min hydrogen flow rate and 25 mL/min air flow rate to the burner). Essentially equivalent relative response factors are demonstrated. This suggests that compound-independent calibration should be possible. In addition, internal standard calibration is attractive because of its applicability to samples of

Table 2.2. Response Factors Relative to Thiophene

Compound	Response Factor
Diethyl sulfide	1.00
Dimethyl disulfide	1.02
Thiophane	1.04
2-Ethylthiophene	1.02
Di-t-butyl disulfide	1.02
n-Octyl mercaptan	0.99
o-Thiocresol	0.99
Benzothiophene	0.97
Diphenyl sulfide	0.99

Table 2.3. Effect of Air Flow Rate on Response Factors[a]

Air Flow Rate (mL/min)	Area of Sulfur (%)		
	Sulfolane	Sulfur (S_8)	Diphenyl Sulfide
15	38.8	31.6	29.6
25	39.1	31.9	29.0
35	38.6	31.9	29.5
Calculated theoretical value	38.1	31.1	30.8

[a]Burner, 775 °C; H_2, 100 mL/min.

varying hydrocarbon composition, i.e., the need to match sample and external standard matrices is eliminated, as will be discussed later.

Although absolute detector response is dependent on operating conditions such as burner temperature and air and hydrogen flow rates to the burner, relative response factors are largely independent of these parameters as long as they are constant throughout an analysis. The influence of air flow rate to the burner and the oxidation state of the sulfur analyte is examined in Table 2.3.

Air flow rates were varied over ranges that might typically be used in flameless SCD, whereas hydrogen flow rate was held constant at 100 mL/min and burner temperature was held fixed at 775 °C. A mixture of nominally 10 ppm S each of sulfolane, elemental sulfur, and diphenyl sulfide in hexane was used in this experiment. The oxidation state of sulfur in these compounds is + 6, 0, and − 2, respectively. The column used in this experiment was a SPB-1, 15 m, 0.53 mm i.d., 0.1 μm film thickness programmed from 40 °C for 1 min to 230 °C for 10 min at 10 °C/min. The injection size was 0.5 μL on-column using a Hewlett Packard 7673 auto-injector. The response of each compound as a normalized percentage of the sulfur in the sample is compared to calculated theoretical values. No attempt was made to assay the purity of the chemicals used in the experiment, although 99% + purities were stated by the manufacturer. Clearly, there is no significant effect of air flow rate on these relative response factors for these compounds even though the sulfur atom is in a different oxidation state in each compound. The ratio of absolute detector response for air flow rates of 15, 25, and 35 mL/min was approximately 6:10:9, respectively.

The analysis of elemental sulfur (S_8 is the major allotrope) in GC/SCD has not been reported previously. Interestingly, in SFC essentially equimolar response was found for all sulfur compounds examined except for elemental sulfur (S_8).[39] This anomaly is not understood but probably is related to the amount of CO_2 entering the burner when used in SFC. In GC, sufficiently thin film columns must be used in order to obtain a sharp peak for the major sulfur allotrope, which will form other allotropes at high temperature. It is interesting that simply turning off hydrogen to the burner produces a very selective detector for elemental sulfur.

It should also be pointed out that spike and recovery experiments demonstrate that the relative response factors of sulfur compounds are not affected by coelution of the sulfur compounds with hydrocarbons.[20,21] Good agreement between total sulfur values obtained by accepted total sulfur methods and methods based on GC/SCD demonstrates this as well.

2.3.2.4. Linearity

In GC, detector linearity to sulfur is typically 3–4 orders of magnitude.[20,21] In SFC, detector linearity to sulfur is about 3 orders of magnitude, with air used for the ozone generator.[39] Curvature at the high end of the calibration curve in SFC may be due to insufficient ozone in the chemiluminescent reaction chamber.

Figure 2.5 shows a plot of calibration curves from eight sulfur-containing compounds in n-hexane; 1 µL splitless injections were made so that the amount of sulfur injected for each component ranged from approximately 1 to 2000 pg. The curves from isopropyl mercaptan and ethyl methyl sulfide exhibit the largest variation in response because these compounds are eluted before the solvent and therefore produce the poorest peak shape. In addition, isopropyl mercaptan is readily oxidized to the corresponding disulfide. Nevertheless, linear least squares analysis results in correlation coefficients greater than 0.9998 for isopropyl mercaptan and ethyl methyl sulfide and greater than 0.9999 for the other compounds. Linear response of 3 orders of magnitude is demonstrated. The least squares fits of the data from each component are nominally equivalent, indicating equivalent response factors for these components. The relative standard deviation (RSD) from the mean of all component response factors is 7.8%.

2.3.2.5. Precision

Flameless SCD response in GC typically exhibits variations of around 2% RSD for stable sulfur compounds in a 1-day period.[20] In SFC, precision of repetitive injections from the system was generally in the range of 2–5% RSD.[39] Figure 2.6 shows a plot of five replicate injections of a standard containing eight sulfur compounds at a 0.5 ppm by weight (ppmw) sulfur of each component, except for dimethyl disulfide, which was present at nominally 1 ppmw as sulfur. This precision of response is comparable to that of an

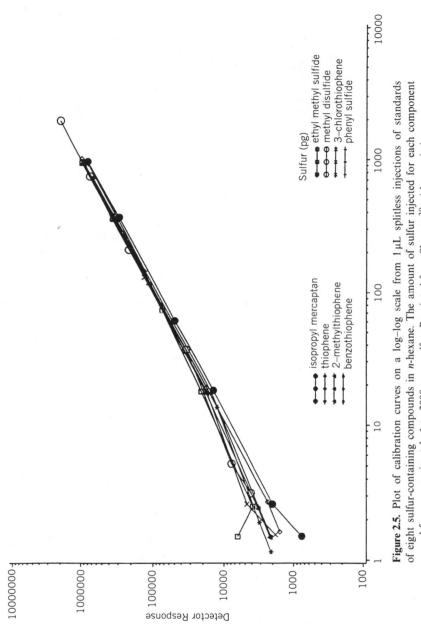

Figure 2.5. Plot of calibration curves on a log–log scale from 1 μL splitless injections of standards of eight sulfur-containing compounds in *n*-hexane. The amount of sulfur injected for each component ranged from approximately 1 to 2000 pg as sulfur. Reprinted from Shearer,[20] with permission.

Sulfur (pg)

isopropyl mercaptan
thiophene
2–methylthiophene
benzothiophene

ethyl methyl sulfide
methyl disulfide
3–chlorothiophene
phenyl sulfide

Detector Response

55

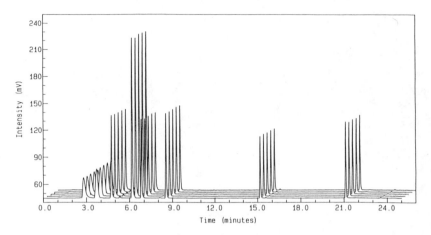

Figure 2.6. Plot of five repetitive injections of a standard of eight sulfur compounds at a 0.5 ppmw sulfur for each component, except for dimethyl disulfide, which was present at nominally 1 ppmw as sulfur. Reprinted from Shearer,[20] with permission.

FID; however, considerable drift in detector response has been observed under some conditions. For instance, raising the temperature of a 30 m, 0.32 mm i.d., 4 μm film thickness SPB-1 column to 360 °C for 1 h between analyses caused a reduction in sensitivity by about 50%.[20] Sensitivity was restored upon reversing the ends of the inner ceramic tube, indicating that surface deactivation occurs on a small zone in the inner tube. During development of flameless SCD, over a dozen ceramic tubes were used and none failed to work on initial installation, although sensitivity usually drifted during the first few hours of operation of new tubes.

The most common cause of poor precision in flameless SCD systems is irreproducible introduction of analytes into the system, resulting from their loss during sampling or analysis (particularly in sample lines from valves and in injection ports). In addition, excessive column and septum bleed causes detector response to change owing to the accumulation of silicon dioxide and other contaminants in the burner, although not to as large a degree as with the flame SCD. Carrier gases should be scrubbed to remove oxygen because oxgen exacerbates column bleed. In addition, column temperatures should be kept as low as possible and, whenever possible, low-bleed columns should be used.

Detector response will vary if hydrogen and air flow rates to the burner are not constant. Pressure regulators should be used upstream of the burner to prevent this. A broken ceramic tube and leaks in the burner will also cause erratic detector response. In general, all of these problems are avoidable.

However, proper quality assurance procedures should be used to validate analytical results. In addition, as will be discussed later, quantitation by the internal standard technique is an effective way to compensate for small drift in detector response, as well as imprecision in injection amounts, sample densities, etc.

2.3.2.6. Other Factors

Some other performance factors related to the operation of the flameless SCD are discussed below. Quenching is a phenomenon common to many selective detectors. Quenching is said to occur when response to an analyte is low due to a negative interference from a matrix component. Quenching in flameless SCD has not been observed to occur with use of normal injection volumes.[20,21] This is in contrast to other selective detectors, e.g., ELCD and FPD, in which a 50% or more loss in analyte signal may occur because of coelution of the sulfur analyte with a major hydrocarbon matrix component.[6]

A minor drawback of the flameless system relative to flame SCD is that an FID signal is not necessarily obtained simultaneously. If an FID signal is desired, one must split the column effluent from a single column or use two columns inserted into the same injection port. In the former case, it is difficult to balance this split between detectors because the flameless system operates under vacuum and an FID operates at ambient pressures. In the latter case, retention times will not match exactly because of differences in the columns and their outlet pressures. In SFC, this problem is virtually nonexistent when the SFC is operated in the downstream mode of operation, in which column pressure and flow control are maintained by a pressure transducer and nozzle downstream of a multiwavelength detector (Figure 2.7). In this configuration, column effluent is easily split to numerous gas phase detectors prior to the multiwavelength detector.

Another advantage of flameless SCD compared to the flame SCD is that the former configuration is easier to operate in that it can be left on continuously. When not in use, the instrument is simply placed in a standby mode (ozone generator off). The reason for this is that almost an order of magnitude less water is produced in the flameless SCD compared to the flame SCD, and this places less demand on the vacuum pump, its oil, and coalescing filter.

The only performance problem experienced during an evaluation of flameless SCD for SFC was detector spiking when column head pressures greater than 250 atm were used.[39] This occurred in the analysis of a nonvolatile residue sample during pressure programming. Apparently, greater CO_2 flow through the restrictor at higher system pressures causes the formation of crystals of CO_2, thereby trapping sulfur-containing solutes. These crystals

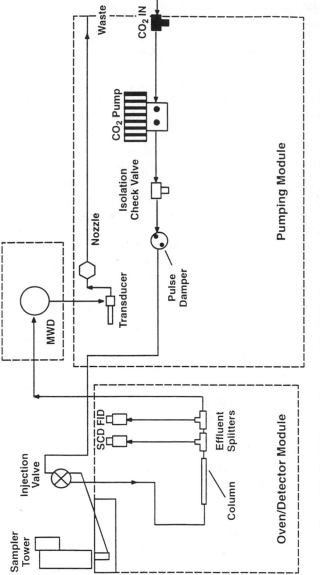

Figure 2.7. Block diagram of a supercritical fluid chromatograph in the downstream mode of operation. System pressure is controlled by a variable restrictor consisting of an automated nozzle and pressure transducer downstream of the column. The majority of column eluent passes through a multiwavelength detector (MWD). A small fraction of the eluent is diverted to an SCD and an FID. Reprinted from Shearer and Skelton,[39] with permission.

then randomly produce spikes upon resublimation. This problem should be correctable through better restrictor heating.

2.4. APPLICATIONS

2.4.1. Gas Chromatography

The analysis of sulfur compounds in petroleum and petrochemical products is important for characterizing crude oils and process streams e.g., by sulfur-simulated distillation), and especially for product quality assurance. This is particularly true for products that come in direct contact with consumers, e.g., solvents for cosmetics and polymers for food packaging, as well as materials that will be processed further by using catalysts that are susceptible to poisoning from certain sulfur-containing compounds. In many instances, it is necessary to measure these compounds at sub-ppm levels. Gas chromatography using flameless sulfur chemiluminescence detection provides many advantages for the analysis of sulfur compounds in petroleum products; this is illustrated by the analysis of sulfur compounds in gasoline.

In the United States, recent legislation at state and federal levels requires reduction of total sulfur levels in fuels such as gasoline and diesel fuel. For instance, all gasoline sold in California will be allowed to contain no more than 40 ppm total sulfur or an average of 30 ppm total sulfur by 1996. This is providing a challenge for refiners because the median level of sulfur in gasoline sold in the United States is approximately 300 ppm sulfur by weight.[41] At the same time, refineries are being required to handle crude slates containing increasing levels of sulfur, compounding the challenge.

In gasoline production, gasoline is blended from various process streams (usually about five or six major streams and some smaller ones depending on the individual refinery) in a dynamic process that is dependent on numerous refinery operations. Optimizing the profitability of a refinery is extremely challenging because desired production levels and distribution of products is dependent on market conditions as well as hardware limitations. Sulfur speciation can aid refiners in understanding how to control sulfur levels in the various process streams in order to meet the legislative requirements.

A GC procedure was developed for the analysis of sulfur compounds in gasoline using an internal standard of 3-chlorothiophene.[21] In the procedure, the internal standard was added to an aliquot of sample prior to GC/SCD analysis, so that the concentration of internal standard was 50–100 ppmw on a sulfur basis. A split ratio of ca. 1:50 is used with the SPB-1 column, and an injection volume of 1 µL was used. Alternatively, on-column injection

Table 2.4. Total Sulfur in Gasolines by Elemental Analysis and Flameless SCD

X-ray Fluorescence (ppmw Sulfur)	Flameless SCD (ppmw Sulfur)	Percent Difference
10	14	40.0
40	41	2.5
160	180	12.5
165	185	12.1
178	187	5.1
189	200	5.8
195	207	6.2
200	184	8.0
240	237	1.3
250	260	4.0
298	294	1.3
430	455	5.8

with an injection volume of 0.1 μL may be used. Individual sulfur constituents were determined directly, and total sulfur was determined by summation of all the peaks excluding the internal standard. (An internal standard should be chosen that does not interfere with sulfur components that are present in the sample.) In some instances, more than one internal standard might prove useful, e.g., to compensate for split discrimination; for example, 2-bromothiophene[19] and diphenyl sulfide.[42] Results from total sulfur analysis of a group of commercial gasolines are given in Table 2.4.

The elemental analysis was performed using an X-ray fluorescence method. Good agreement was achieved between the X-ray and the GC methods. In most cases, results agree to within 10% relative. The analysis of the gasoline containing approximately 10 ppmw sulfur is near the detection limit of the elemental analysis method, and greater discrepancy between results is expected. Similar results were reported from the analysis of light petroleum streams in a comparison of elemental analysis and conventional SCD analysis.[19] As pointed out previously, such results demonstrate that coeluting hydrocarbons do not significantly affect detector response to sulfur. Very good precision of 1–2% RSD was typically obtained for replicates of total sulfur and individual sulfur compounds in gasoline by the GC/flameless SCD method.

The ability to speciate the sulfur compounds is an advantage of the GC method over elemental analysis. The retention times for a series of thiophenes present in light petroleum streams were given previously.[19] Figure 2.8 shows

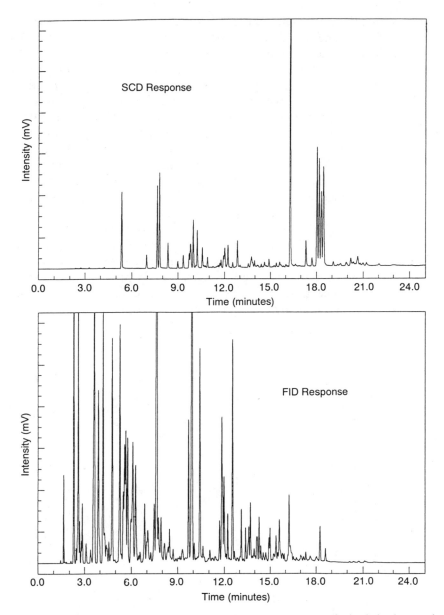

Figure 2.8. Comparison of flameless SCD and FID chromatograms obtained simultaneously from an injection from a gasoline sample containing approximately 300 ppmw sulfur. GC column: SPB-1, 30 m, 0.32 mm i.d., 4 μm film thickness. Conditions: 1 μl injection split 1:50, oven temperature 40 °C for 3 min to 275 °C at 10 °C/min. Reprinted from Shearer et al.,[21] with permission.

flameless SCD and FID chromatograms obtained simultaneously from the analysis of a gasoline containing ca. 300 ppmw sulfur. An outlet splitter was used for simultaneous SCD and FID detection. An SPB-1 column used for this analysis was held at 40 °C for 3 min and temperature programmed at 10 °C/min to 275 °C. Figure 2.8 readily demonstrates the excellent selectivity of the detector. This selectivity allows determination of sub-ppm levels of sulfur compounds even in the presence of percent levels of coeluting hydrocarbons. Conversely, complete or nearly complete chromatographic resolution would be required for sub-ppm level measurements by GC/AED or FPD.

Flameless SCD has also been used successfully for the measurement of a common odorant (ethyl mercaptan) that is used in liquefied petroleum gas (LPG).[21] The system has been in operation in the quality assurance laboratory of a Shell refinery and used daily for almost 2 years with minimal maintenance. In addition, through the use of flameless SCD, precision of the analysis was improved by a factor of almost 6, relative to flame SCD, and the detector has been easier to operate. Figure 2.9 is a control chart from the analysis of a quality assurance standard of ethyl mercaptan in propane at a level of 2.5 lb/10,000 gal. The range between the upper and lower control limits corresponds to approximately 3.5 ppmw ethyl mercaptan. Figure 2.10 is a

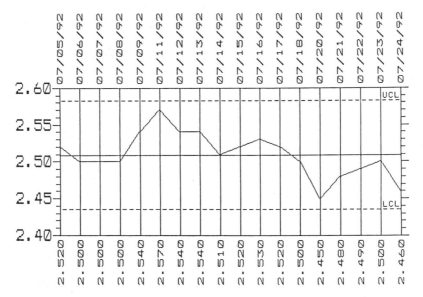

Figure 2.9. Control chart for a quality assurance standard of ethyl mercaptan in propane present at a concentration of approximately 2.5 lb per 10,000 gal of liquid (60 ppmw). The range between the upper and lower control limits is approximately ± 1.8 ppmw. Reprinted from Shearer et al.,[21] with permission.

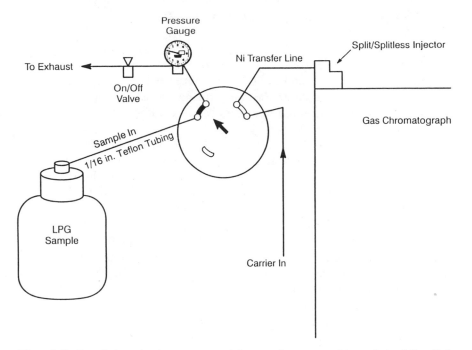

Figure 2.10. Sample introduction system used for gas chromatographic analysis of liquefied petroleum gases (LPGs). Reprinted from *et al.*,[21] with permission.

diagram of the LPG sample introduction system. The line between the liquid sampling valve and the inlet to the injection port is made of electropolished nickel, except for a 2 cm section of the original stainless steel tubing at the injector.

In GC, it is possible to identify and quantitate nearly all of the sulfur compounds found in petrochemical products that have boiling points below 80 °C. The number of possible sulfur-containing isomers increase exponentially above that temperature range; thus it is usually only possible to identify major components unambiguously. For instance, only 14 sulfur-containing compounds that boil below 80 °C were identified in crude oil, but 17 sulfur-containing compounds that boil in the range between 84 °C (boiling point of thiophene) and 113 °C (boiling point of 2-methylthiophene) were identified in crude oil.[43] Of course, it is not always necessary to measure individual compounds. It is often desirable to measure only specific sulfur compound classes, such as mercaptans, sulfides, disulfides, and thiophenes. Class-type separation such as this are probably more easily, accomplished by LC and SFC.

2.4.2. Supercritical Fluid Chromatography

SFC is more compatible with gas phase detectors than is LC, and this is a great advantage of SFC over LC. Sulfur analysis by SFC with plasma detection,[44,45] FPD,[46,47] and SCD[48-51] has recently been described. Among these, the flame SCD has shown the best performance in terms of detectability, selectivity, linearity and uniform sulfur response, although each technique may possess some advantages.

SFC using flameless SCD also works well for the analysis of sulfur compounds in petroleum products. The flameless SCD is more sensitive and selective than are other sulfur selective detectors used in SFC.[39] It also produces a linear and nearly equimolar response to sulfur, and it is especially applicable to the analysis of sulfur compounds in complex matrices such as those found in the petroleum industry. The effect of CO_2 on detector response was found to be negligible at decompressed flow rates under 10 mL/min.[39] Therefore, no SFC-specific optimization of the detector was necessary. Flameless SCD was easy to implement in SFC, and application was straightforward while providing very valuable information.

For example, Figure 2.11 shows an isometric plot of two sulfur chromatograms (one of a feed sample, and one of a product sample from a catalytic hydrocracking unit) obtained by SFC with flameless SCD.[39] THE SFC was operated in the downstream mode of operation, as shown in Figure 2.6. Two frit restrictors were used as effluent splitters to divide a portion of the column effluent between a flameless SCD and an FID. The flow rate of CO_2 through

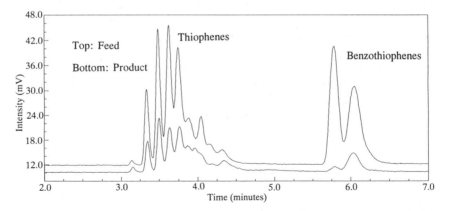

Figure 2.11. Comparison of a hydrocracking unit feed (top chromatogram) and product (bottom chromatogram) using flameless SCD with SFC. A silica column operated at 55 °C and 120 atm pressure with a 2.5 mL/min CO_2 flow rate; 5 μL injection was used to separate thiophenes from benzothiophenes. Reprinted from Shearer and Skelton,[39] with permission.

each restrictor was approximately 2.5 mL/min at STP (ca. 0.3% of the total column effluent). The column consisted of two serially coupled silica columns, each 250 mm × 4.6 mm, of 5 μm particle size. The SFC was operated in the constant flow mode of 2.5 mL/min at a constant pressure of 120 atm and temperature of 55 °C. The injection volume was 5 μL. The top chromatogram is a plot of the feed containing approximately 0.1 wt% sulfur. The bottom plot is that of the product containing approximately 300 ppm sulfur. The sulfur compounds in the samples are separated into groups of thiophenes and benzothiophenes. Had thiophenols been present, they would have been eluted between these two sulfur compound classes.

FID and UV (254 nm) chromatograms of the feed sample, obtained simultaneously with the sulfur chromatogram, are shown in Figure 2.12. The FID chromatogram illustrates the chromatographic selectivity of the separation in that the saturates and olefins of carbon number less than C_{22} are eluted by 3 min, whereas the aromatics (one and two rings) are eluted after 3 min. The UV chromatogram illustrates the strong response to aromatic compounds and further demonstrates the power of this combination of chromatographic and detector selectivity.

At present, the use of modifier phase modifiers in SFC has not been evaluated with flameless SCD, but the flame SCD has been shown to be

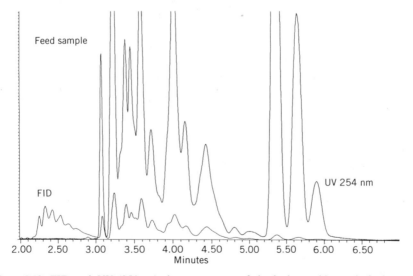

Figure 2.12. FID and UV (254 nm) chromatograms of the hydrocracking unit feed sample obtained simultaneously with the flameless SCD chromatogram shown in Figure 2.11. The FID chromatogram is the upper trace in the plot prior to 3 min, and it is the lower trace thereafter. Reprinted from Shearer and Skelton,[39] with permission.

compatible with methanol-modified carbon dioxide as the mobile phase.[48] The use of mobile phase modifiers in SFC with flameless SCD should be the object of further study.

2.4.3. Liquid Chromatography

The flame SCD has also been interfaced successfully with capillary liquid chromatography,[52] and in a review of its use in GC, SFC, and LC, Howard and Taylor concluded that the flame SCD is an extremely versatile detector for these techniques and clearly surpasses other sulfur selective detectors employed in those areas.[53] Flameless SCD should demonstrate additional advantages when it is employed in LC, and it may also prove useful for capillary zone electrophoresis.

2.5. CONCLUDING REMARKS

The development of flameless SCD provides an excellent means to measure sulfur-containing compounds by GC, SFC, and possibly LC as well. The major reasons for its success are its excellent sensitivity and its ability to discriminate against interfering compounds, i.e., its selectivity. The flameless SCD compares most favorably with other sulfur-selective detectors in all aspects of its performance as shown in Table 2.5. The values given in the table are from manufacturer's specifications or from the scientific literature.[5-9,16-21] Of course, it is difficult to cover all aspects of these detectors, but in most instances flameless SCD should be considered the sulfur-selective detector of choice. Nevertheless, each detector may possess certain advantages for particular applications. For instance, the FPD is small and relatively

Table 2.5. Comparison of Commercially Available Sulfur-Selective Detectors

Detector	Minimum Detectability (pg S/s)	Selectivity (S/C)	Linear Response	Other Characteristics
FPD	2–50	10^4–10^5	No	Quenching, inexpensive
ELCD	2–50	10^4–10^5	Yes	Difficult to use, inexpensive
AED	1–2	1.5×10^5	Yes	Expensive, skilled analysts, multielement detection
Flame SCD	0.5–5	$> 10^7$	Yes	Skilled operators
Flameless SCD	0.05–0.5	$> 10^7$	Yes	Easy to operate

inexpensive and can be used adequately for quantitation of sulfur compounds in some samples. It might be good for use with protable instruments and where interferences do not greatly affect its accuracy. On the other hand, the AED is large and expensive and requires skilled operators because of its complexity. but it is also a very powerful multiple element detector that serves as a useful analytical tool. As technology continues to advance, these detectors may become better and new detectors may be developed. For example, a pulsed flame technique may improve both the selectivity, sensitivity, and perhaps some other aspects of the FPD, but it still will suffer from quenching and nonlinearity.[54] In any case, it is a challenge for analytical chemists and engineers to develop new instrumental methods and to improve upon old ones, and continued developments are foreseeable in the area of flameless SCD.

Interestingly, the flame SCD was originally developed as a total sulfur analyzer for measuring sulfur compounds in the atmosphere.[14,28] The flameless SCD demonstrates very good performance for this.[55] It also shows promise as a total sulfur analyzer for measuring low levels of sulfur compounds in process streams.[56]

Safety and environmental concerns compel us toward further improvements in all aspects of our analytical methods, sample preparation, separation, and detection. Therefore, the practicing chromatographer will continue to be asked to measure more components at lower levels and in more complex matrices than ever before. In the area of chromatography related to sulfur chemiluminescence detection, further research should be undertaken to better understand all factors affecting the performance of the detector with the objective of improving its performance even further. Research into coupling flameless SCD with LC, with SFC, using modified CO_2, and with capillary zone electrophoresis should proceed in order to meet current and future analytical needs.

REFERENCES

1. H. V. Drushel, *J. Chromatogr. Sci.* **21**, 375 (1983).
2. ASTM D 130, *Annual Book of ASTM Standards*, Vol. 05.01. ASTM, Philadelphia, 1988.
3. A. M. Patterson, *Chem. Eng. News* **43**(36), 3681 (1953).
4. S. S. Brody and J. E. Chaney, *J. Gas Chromatogr.* **4**, 42 (1966).
5. S. O. Farwell and C. J. Barinaga, *J. Chromatogr. Sci.* **24**, 483 (1986).
6. M. Dressler, *Selective Gas Chromatographic Detectors*, J. Chromatogr. Libr., Vol. 36. Elsevier; Amsterdam, 1986.
7. R. C. Hall, *J. Chromatogr. Sci.* **12**, 152 (1974).

8. P. C. Uden, Y. Young, T. Wang, and Z. Cheng, *J. Chromatogr.* **468**, 319 (1989).

9. B. D. Quimby and J. J. Sullivan, *Anal. Chem.* **61**, 1027 (1990).

10. J. K. Nelson, R. H. Getty, and J. W. Birks, *Anal. Chem.* **55**, 1767 (1983).

11. S. A. Nyarady, R. M. Barkley, and R. E. Sievers, *Anal. Chem.* **57**, 2074 (1985).

12. W. Bruening and F. J. M. Concha, *J. Chromatogr.* **142**, 191 (1971).

13. J. S. Gaffney, D. J. Spandau, T. J. Kelly, and R. L. Tanner, *J. Chromatogr.* **347**, 121 (1985).

14. R. L. Benner and D. H. Stedman, *Anal. Chem.* **61**, 1268 (1989).

15. J. W. Birks, *Chemiluminescence and Photochemical Reaction Detection in Chromatography.* VCH Publishers, New York, 1989.

16. S. E. Eckert-Tilotta, S. B. Hawthorne, and D. J. Miller, *J. Chromatogr.* **591**, 313 (1992).

17. R. L. Shearer, D. L. O'Neal, R. Rios, and M. D. Baker, *J. Chromatogr. Sci.* **28**, 24 (1990).

18. R. S. Hutte, N. G. Johansen, and M. F. Legier, *J. High Resolut. Chromatogr.* **13**, 421 (1990).

19. C. Birbal and F. Di Sanzo, *J. Chromatogr.* **589**, 271 (1992).

20. R. L. Shearer, *Anal. Chem.* **64**, 2192 (1992).

21. R. L. Shearer, E. B. Poole, and J. B. Nowalk, *J. Chromatogr. Sci.* **31**, 21 (1993).

22. J. C. Kramlich, The fate and behavior of fuel sulfur in combustion systems. Ph.D. Thesis, Washington State University, Pullman (1980).

23. G. M. Johnson, C. J. Matthews, M. Y. Smith, and D. J. Williams, *Combust. Flame* **15**, 211 (1970).

24. A. S. Kallend, *Combust. Flame* **13**, 324 (1969).

25. A. Demerdache and T. M. Sugden, in *The Mechanism of Corrosion by Fuel Impurities* (H. R. Johnson and D. J. Littler, eds.) Butterworth, London, 1963.

26. C. P. Fenimore and G. W. Jones, *J. Phys. Chem.* **69**, 3593 (1965).

27. E. M. Bulewicz and T. M. Sugden, *Proc. R. Soc. London, Ser. A* **277**, 143 (1964).

28. R. L. Benner, Development of the sulfur chemiluminescence detector. Ph.D. Thesis, University of Denver, Denver, CO (1991).

29. D. J. W. Kendall, J. J. A. O'Brian, J. J. Sloan, and R. G. MacDonald, *Chem. Phys. Lett.* **110**, 183 (1984).

30. C. J. Halstead and B. A. Thrush, *Proc. R. Soc. London* **295**, 380 (1966).

31. H. R. Martin and R. J. Glinski, *Appl. Spectrosc.* **46**, 948 (1992).

32. H. Akimoto, B. J. Finlayson, and J. N. Pitts, *Chem. Phys. Lett.* **12**, 199 (1971).

33. M. A. A. Clyne, C. J. Halstead, and B. A. Thrush, *Proc. R. Soc. London* **295**, 355 (1966).

34. W. B. DeMore, J. O. Molina, S. P. Sander, D. M. Golden, R. F. Hampson, M. J. Kurylo, C. J. Howard, and A. R. Ravishankara, *JPL Publ.* **87-41** (1987).

35. R. D. Cadle and M. Ledford, *Air, Water Pollut.* **10**, 25 (1966).

36. J. M. Hales, J. O. Wilkes, and J. L. York, *Atmos. Environ.* **3**, 657 (1969).

37. J. G. Burr, *Chemi- and Bioluminescence.* Dekker, New York, 1985.

38. P. W. Schenk and R. Steudel, *Angew. Chem., Int. Ed. Engl.* **4**, 402 (1965).

39. R. L. Shearer and R. J. Skelton, *J. High Resolut. Chromatogr* (submitted for publication) (1993).

40. R. L. Grob, *Modern Practice of Gas Chromatography.* Wiley, New York, 1985.

41. Motor Vehicle Manufacturers Association, Survey Data (1990).

42. G. D. Dupré and H. V. Drushel, *Pittsburgh Conf.*, Chicago, *1991*, Pap. No. 575 (1991).

43. American Petroleum Institute, Project 44. API, New York, 1959.

44. D. R. Luffer, L. J. Galante, P. A. David, M. Novotny, and G. M. Hieftje, *Anal. Chem.* **60**, 1365 (1988).

45. R. J. Skelton, P. B. Farnsworth, K. E. Markides, and M. L. Lee, *Anal. Chem.* **61**, 1813 (1989).

46. S. V. Olesik, L. A. Pekay, and E. A. Paliwoda, *Anal. Chem.* **61**, 58 (1989).

47. L. A. Pekay and S. V. Olesik, *Anal. Chem.* **61**, 2616 (1989).

48. H.-C. K. Chang and L. T. Taylor, *J. Chromatogr.* **517**, 461 (1990).

49. L. A. Pekay and S. V. Olesik, *J. Microcolumn Sep.* **2**, 270 (1990).

50. W. F. Sue, Z. X. Zhao, and M. L. Lee, *Chromatographia* **33**, 507 (1992).

51. A. L. Howard and L. T. Taylor, *Anal. Chem.* **65**, 724 (1993).

52. H.-C. K. Chang and L. T. Taylor, *Anal. Chem.* **63**, 486 (1991).

53. A. L. Howard and L. T. Taylor, *J. High Resolut. Chromatogr.* **14**, 785 (1991).

54. E. Atar, S. Cheskis, and A. Amirav, *Anal. Chem.* **63**, 2064 (1991).

55. J. R. Burdge and S. O. Farwell, *Pittsburgh Conf.*, Atlanta, GA, *1993*, Pap. No. 628 (1993).

56. J. D. Ray, N. J. Johanson, and R. S. Hutte, *Pittsburgh Conf.* Atlanta, GA, *1993*, Pap. No. 607 (1993).

CHAPTER

3

SELECTIVE DETECTION TECHNIQUES BASED ON NITRIC OXIDE/OZONE CHEMILUMINESCENCE

ANDREW J. DUNHAM and ROBERT E. SIEVERS

Department of Chemistry and Biochemistry, Cooperative Institute for Research in Environmental Sciences (CIRES), and Global Change and Environmental Quality Program, University of Colorado at Boulder, Boulder, Colorado 80309

Chemiluminescence is the emission of light from excited species formed in a chemical reaction. The application of chemiluminescence to sensitive analyses in complex samples is advantageous for three reasons: chemiluminescence is observed as a single against a dark, low-noise background; the reaction generating emission occurs between a limited number of compounds; and the measurement of specific emissions can be selected using optical filters. Perhaps the most widely used gas phase chemiluminescent reaction for analysis is between nitric oxide and ozone. This chapter will review the detectors and techniques based on chemiluminescence from this reaction, which allows highly sensitive and selective measurement of nitric oxide and of precursors of oxides of nitrogen.

3.1. INTRODUCTION

3.1.1. History

Chemiluminescence from electric discharge air afterglow in vacuum tubes was first observed in the mid-nineteenth century. In 1910, Strutt determined the source of the emission: the chemiluminescent reaction between nitric oxide (NO) and ozone (O_3).[1] Additional experimentation in 1949 by Tanaka and Shimazu[2] and 1959 by Greaves and Garvin[3] demonstrated that the reaction between O_3 and NO generated a red-yellow emission whereas reac-

Selective Detectors, edited by Robert E. Sievers. Chemical Analysis Series, Vol. 131.
ISBN 0-471-01343-9 © 1995 John Wiley & Sons, Inc.

tion of O and NO yielded a green-yellow emission. Kinetic and mechanistic aspects of the NO + O$_3$ reaction were presented in detail by Clyne et al. in 1964[4] and Clough and Thrush in 1967[5]; the full spectrum of the emission, 700–3200 nm, with a maximum around 1200 nm, was reported in 1966.[5] Four years later Fontijn et al., reported the construction of an instrument for chemiluminescent measurement of nitric oxide in air.[6] The instrument was developed to show the potential of chemiluminescence from reactions of O$_3$ or O for measuring various atmospheric pollutants. A fundamental detector design based upon a chemiluminescence reaction with sufficient exothermicity to provide excellent sensitivity[6] stimulated the development of new analytical techniques and instruments. The ability to analyze sensitively a compound or group of compounds based upon the generation of NO from many kinds of samples has opened new doors to trace analyses of a host of samples with complex matrices.

3.1.2. Principle of Operation

The process of NO chemiluminescence can be illustrated with a few simple reactions. The bimolecular reaction between NO and O$_3$ yields an electronically excited NO$_2$ molecule [see Reaction (1)–(4)]. Relaxation from low-lying vibrational levels of the excited electronic state to the ground state generates emission of radiation over a spectrum from 600 to 3200 nm, with a maximum around 1200 nm (see Figure 3.1).[5]

$$NO(g) + O_3(g) \xrightarrow{k_1} NO_2^* + O_2 \tag{1}$$

$$NO(g) + O_3(g) \xrightarrow{k_2} NO_2 + O_2 \tag{2}$$

$$NO_2^* \xrightarrow{k_3} NO_2 + h\nu \tag{3}$$

$$NO_2^* + M \xrightarrow{k_4} NO_2 + M \tag{4}$$

The reaction was studied closely in a flow system at pressures between 0.8 and 3.0 torr and temperatures from 216 to 322 K; the overall rate constant determined by Clyne et al. is $5.7 \times 10^{11} \exp(-2460 \pm 150/RT) \, \text{cm}^3 \, \text{mol}^{-1} \, \text{s}^{-1}$.[4] The overall reaction:

$$NO(g) + O_3(g) = NO_2^*(g) + O_2(g) \tag{5}$$

is exothermic by 48 kcal/mol.[7] The effect reaction chamber pressure has on

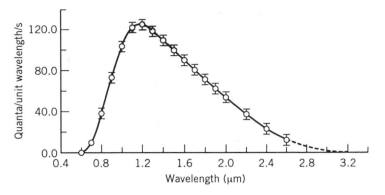

Figure 3.1. Spectrum of NO_2^* emission.

the chemiluminescent response is apparent from Reaction (4). Half-quenching pressures, the pressures at which half of the NO_2^* formed emits light and half is collisionally quenched, are 13.0, 8.3, 5.2, and 3.7 µm Hg for Ar, O_2, NO, and NO_2, respectively.[5] These instruments are generally operated at pressures less than 50 torr. The resulting emission is monitored with a photomultiplier tube (PMT) sensitive to red wavelengths; a trialkali-type PMT has been used extensively.[6] The temperature of the PMT is reduced with thermoelectric cooling to realize the maximum possible sensitivity by reducing the dark current. However, the bulk of the NO_2^* emission spectrum falls outside of the sensitive range of the PMT (see Figure 3.2). In addition, the quantum efficiency of the PMT for the part of the NO_2^* emission detected is much less than unity. Taking into account the quantum efficiency and losses, such as from reflection, for a NO chemiluminescence detector constructed with a suitable PMT and optical filter to cut off wavelengths less than 620 nm, one can estimate the overall photon collection efficiency to be 0.00002.[8] Nevertheless, the detection of NO can be accomplished with both great sensitivity and selectivity.

The observable photon flux, $h\nu_R$, is given by

$$h\nu_R = \left(\frac{k_1}{k_1 + k_2}\right)\left(\frac{k_3}{k_4}\right)\left(\frac{Q_{sample}}{Q_{net}}\right)\rho\mu_{NO} \tag{6}$$

where k_1 through k_4 are the rate constants from Reactions (1)–(4); Q_{sample} and Q_{net} are the sample and total flow rates, respectively; ρ is the volumetric pumping rate of the detector vacuum pump; and μ_{NO} is the mixing ratio of nitric oxide.[9,10] For instruments in which the sample flow rate and total flow rate are approximately equal, such as those used in the measurement

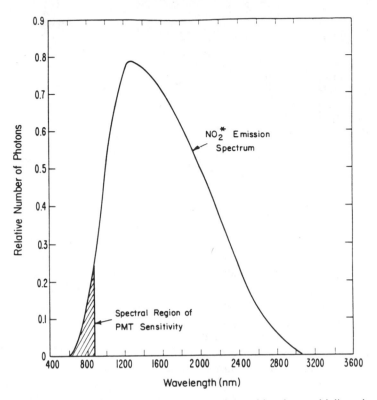

Figure 3.2. Portion of NO$_2^*$ emission spectrum detected by photomultiplier tube.

of nitric oxide in the atmosphere, $Q_{sample}/Q_{net} \approx 1.$[9] Since k_1 through k_4 and ρ are constants, hv_R is proportional to μ_{NO}. When the instrument is used as a chromatographic detector, however, the sample flow rate may be much smaller than the total flow rate (such as the flow from a capillary gas chromatograph). In order for hv_R to be proportional to μ_{NO}, Q_{sample} and Q_{net} must be held constant. The flows are controlled in practice by allowing ozone and sample gas into the reaction chamber through flow controllers. The reduced pressure in the reaction chamber is provided by a vacuum pump; an ozone scrubber is placed before the vacuum pump to protect the oil from degradation by excess ozone from the reaction chamber.

3.1.3. Summary of Instrument Development

Several techniques for a variety of applications have been developed based upon NO chemiluminescence. The group analysis of *N*-nitroso compounds in food was reported in 1973 and 1974.[11,12] NO chemiluminescence was applied to organonitrogen compounds in 1977 in a patent awarded to Parks and Marietta.[13] The conversion of nitrate and nitrite ions in aqueous samples to NO was presented in 1980.[14,15] The application of NO chemiluminescence to the measurement of NO in aqueous solution followed.[16] NO chemiluminescence detection was applied to the atmospheric analysis of NO, NO₂, and NH₃ in 1973,[17] and later broadened to reactive, odd-nitrogen compounds with the development of a technique for reaction of these compounds with carbon monoxide over a gold catalyst to yield NO for detection.[8,18,19] Surrogate NO produced via the oxidation of compounds by NO₂ was reported by Nyarady and Sievers in 1985[20] for use as a gas chromatographic detector.[21] The development and applications of these techniques will be described in detail below.

3.2. ANALYSES OF AQUEOUS NO, NITRITE ION, AND NITRATE ION BY NO CHEMILUMINESCENCE

3.2.1. Introductory Remarks

The quantitative analysis of nitrogen compounds in water facilitates our understanding of the nitrogen cycle and is an indicator of water quality. Of particular interest from a human health perspective are the concentrations of nitrate or nitrite ions in water or food samples. Salts of nitrite or nitrate are either directly toxic or are precursors to toxic compounds or suspected carcinogens. Conventionally, nitrate and nitrite ion have been measured by a number of techniques, including spectrophotometry[22] and chromatography.[23–30] However, in a complex sample matrix, turbidity or color can interfere with the measurements by spectrophotometric methods, while the chromatographic methods have in the past lacked the sensitivity necessary to detect trace quantities.

3.2.2. Analysis of Nitrite and Nitrate Ions in Aqueous Samples

Reduction of nitrate and nitrite ions to NO with subsequent chemiluminescence detection provides a sensitive method for analysis that avoids interferences due to other constituents in the sample matrix. Since the detection and measurement occurs in the gas phase, the color and turbidity of the original aqueous sample from which the gaseous NO has been removed do not interfere with the analysis. The reduction of nitrate ion and nitrite ion, followed by chemiluminescence detection, was first reported by Cox in 1980.[15] The apparatus (see Figure 3.3) is composed of a pair of impingers. The first

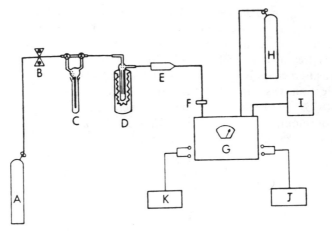

Figure 3.3. Apparatus for reduction of nitrate and nitrite ions to nitric oxide by ozone chemilu-minescence: A, He tank; B, needle valve; C, impinger containing reducing agent; D, cold trap; E, vapor scrubber, F, particle filter; G, NO$_x$ analyzer; H, oxygen; I, vacuum pump; J, strip chart recorder; K, integrator.

is a reaction flask into which the sample is introduced to mix with the appropriate reducing agents. The second impinger acts as a cold trap for water vapor removal. Helium sweeps the products of the reduction through the impingers to an evacuated chamber for reaction with ozone.

Two separate reagent systems are used for the conversion of nitrate ion and nitrite ion to nitric oxide. Both nitrate and nitrite ion are reduced to nitric oxide using 5 mL of H$_2$SO$_4$ with 1 mL of 4% ferrous ammonium sulfate and 1 mL of 2% ammonium molybdate, or 5 mL of H$_2$SO$_4$ and 1 mL of 5% titanium trichloride. In an independent measurement of a different aliquot of the same sample, selective reduction of only nitrite ion to nitric oxide is achieved with 1 mL of 0.2 M NaI and 3 mL of acetic acid. The nitrate ion concentration is then calculated by difference. The detection limits for nitrate ion and nitrite ion are 0.05 µg/L in a 20 mL sample and 5 µg/L in a 5 mL sample, respectively. The selectivity of this method over other compounds is excellent. Of the several N-nitroso compounds tested only N-nitrosodi-phenylamine produced NO under the reduction conditions; alkyl nitrites or other compounds which decompose under acidic conditions to form nitrite will also interfere, if present. This method has been applied to the analysis of nitrate ion and nitrite ion in seawater.[31]

3.2.3. Analysis of NO and Nitrite Ion in Seawater

The analysis of nitrite ion and nitric oxide in seawater was reported in 1980 by Zafiriou et al.[14] Nitric oxide in seawater is measured directly by sparging water below the sea surface with air or N$_2$. The resulting bubbles are collected in a poly(ethylene) jar; the collection vessel is directly sampled by the NO chemiluminescence detector.[14] This method is sensitive to $< 10^{-13}$ M NO in water.[16] The analysis of nitrite ion in the sample is achieved via photolysis of the sample with ultraviolet (UV) light, and the resulting NO is sparged from the water into a chemiluminescence detector. The sample is placed in a large borosilicate jar and exposed to solar UV radiation or UV radiation from a Xe-arc lamp with a Pyrex filter. The Xe-arc lamp with Pyrex filter approximates the sea level sunlight spectrum. NO produced by the photolysis is stripped from the solution using high-purity compressed air or nitrogen. Measurements were reported for both *in situ* and laboratory analysis of NO(*aq*) in seawater.[14,16]

3.2.4. Analysis of Nitrate Ion in Atmospheric Particles

The concentration of nitrate ion in tropospheric aerosol particles is one measure of anthropogenic pollution. The sensitivity and selectivity of NO chemiluminescence has led to the development of another method for this measurement.[32] The apparatus employed dual impingers as previously described, but 90% phosphoric acid heated to 100 °C is used instead of ferrous or molybdate reducing agents. An injector is employed for sample introduction. The optimum response is not temperature dependent over the range of 80–110 °C; however, the half-width of the observed peak decreased with increasing temperature. Since the generation of water vapor, which is deleterious to sensitive measurements, also increased with temperature, 100 °C was chosen as the optimum working temperature. The reduction of nitrate ion in samples of atmospheric aerosols and gas phase nitric acid collected in downtown Tokyo was accomplished. Atmospheric aerosols were collected on a Teflon filter, while gas phase nitric acid was trapped on a polyamide filter. Nitrate ion was extracted from the polyamide filter with water; extraction from the Teflon filter required ethanol pretreatment before aqueous extraction. The method was linear from 1 to 1000 ng of NO$_3^-$, with a detection limit of 1 ng of NO$_3^-$ mL for a 5 mL sample. In order to gather sufficient aerosol sample for analysis, the researchers drew air for 60 min through the polyamide filter.

3.2.5. Flow Injection Analysis Techniques for Nitrite Ion in Aqueous Samples

Flow injection analysis methods for nitrite ion based on liquid phase conversion to NO with gas phase chemiluminescence detection have been reported.[33,34] The conversion in the first method[33] is accomplished by reaction of nitrite ion in the sample in a flowing stream of 0.1 M H$_3$PO$_4$

without a reducing agent. The protonation of nitrite ion yields nitrous acid (HNO_2); the pK_a of this acid at 25 °C is approximately 4.3. At room temperature nitrous acid disproportionates to NO and NO_3^-:

$$3NO_2^- + 3H^+ \rightleftharpoons 3HNO_2 \rightleftharpoons 2NO + NO_3^- + H_3O^+ \qquad (7)$$

Disproportionation converts only two-thirds of the available nitrite ion to NO. Nitric oxide produced in the acidification and subsequent dispro-portionation is swept from the liquid phase to the gas phase via nebulization with argon (Ar). Excess water is collected in a water-cooled condenser and removed by suction with a pump. A detection limit of 2 pmol of $NaNO_2$ was reported.[33] However, the method showed limited linearity and a high blank response. Potential interferences include alkyl thionitrite and alkyl nitrites, which yielded equimolar responses with nitrite ion, whereas *N*-nitroso compounds showed no significant response.

The second flow injection analysis method[34] uses a semipermeable mem-brane to strip the NO formed in the liquid phase reduction to the gas phase for detection.[34] The sample solution is the liquid carrier in the system and is combined with a reducing agent stream [1.0 M NaI in 0.18 M $H_2SO_4(aq)$]

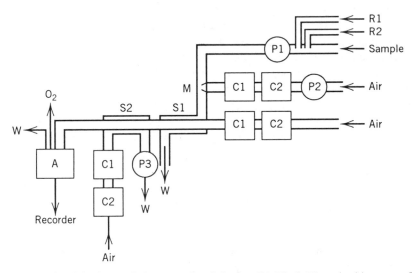

Figure 3.4. Flow injection analysis system for nitrite ion: P1, P2, & P3, peristaltic pumps; R1, 1.0 M potassium iodide (*aq*); R2, 0.18 M sulfuric acid (*aq*); C1 & C2; silica gel and activated charcoal columns, respectively; S1, porous PTFE tubing for NO separation from liquid phase; S2, Nafion tubing for removal of water; A, chemiluminescence reaction chamber; W, waste, M mixing tee.

at 0.8 mL/min (see Figure 3.4). The sample is then pumped at 8 mL/min into the separation unit. Air bubbles are introduced to the liquid stream here to facilitate stripping of NO through the microporous, semipermeable poly(tetrafluoroethylene) (PTFE) tubing. The gas separation unit is 1200 mm long (total) and consists of two sets of two concentric tubes: nonporous PTFE tubing around microporous PTFE (600 mm long) tubing, followed by nonporous PTFE tubing around Nafion (600 mm) tubing. The NO produced in the liquid phase is stripped by the air bubbles through the permeable PTFE tubing to the gas phase. The gas stream then passes through the Nafion tubing, where water vapor is removed. The dried NO-containing gas stream flows to the detector, where the ozone chemiluminescent reaction can be observed. A stable signal is obtained after approximately 1.5 min, requiring 12 mL of sample. The method was linear over the range of 10.0 ppb to 5.0 ppm, with a detection limit [signal-to-noise ratio $(S/N) = 3$] of 0.9 ppb, with a relative standard deviation (RSD) at 100 ppb of 0.7% $(n = 5)$.

A novel technique has been developed[35] in which small aliquots of a sample (5–160 µL) are injected into a flowing reagent stream (see Figure 3.5). The reducing reagent is 0.1 M sodium iodide in aqueous 0.1 M phosphoric acid, delivered with a high-performance liquid chromatography (HPLC) pump at 1.0 mL/min. The sample is introduced through a fixed-volume injection valve and is carried through a short section of inert, impermeable capillary tubing (e.g., fused silica) to allow the reduction to occur. The stream then passes through a 15 cm long section of semipermeable PTFE tubing (Gore-tex) that contains a single-bead string reactor (SBSR)[36,37] constructed of nylon fishing line. The SBSR is used to break up laminar flow in the semipermeable tubing, increasing contact between the sample plug and the walls of the tube. Nitric oxide produced in the reduction passes through the membrane and is swept to the gas phase for detection by ozone chemilumi-

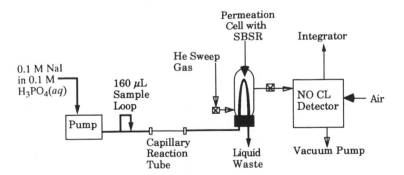

Figure 3.5. Flow injection analysis system for nitrite from small (microliter) samples.

Table 3.1. Comparison of Measurements of Nitrite by Chemiluminscence vs. UV Absorbance

Sample	Nitrite Concentration $\pm$ 1s[a]	
	Diazotization/UV	FIA/CL
Saliva (ppm nitrite)	35.0 $\pm$ 0.8	34.5 $\pm$ 0.9
Hot dog extract (μg nitrite/g food)	17.7 $\pm$ 0.4	19.8 $\pm$ 0.2
Lettuce extract (μg nitrite/g food)	Below detection limit	0.19 $\pm$ 0.03

[a]Here s = sample standard deviation.

nescence. The detector used for these measurements was a Sievers Instruments Co. Model 207 RCD™, in which the flow system for redox chemiluminescence detection was bypassed.

The detection limit for a 100 μL sample is 0.04 ppb nitrite ion (where signal/σ_{RMS} = 3.29). The sampling rate is dependent on the volume of sample injected (e.g. 5–160 μL); between 60 and 180 samples can be analysed per hour. In addition, the conversion reaction shows no temperature dependence over the range of 20–70 °C. Results from nitrite analysis on samples of human saliva and food extracts by this method and the diazotization reaction/spectrophotometric method[22] are shown in Table 3.1. The concentration of nitrite ion present in the lettuce extract was below the detection limit of the diazotization reaction/spectrophotometric method; however, the flow injection analysis (FIA) method did detect a low concentration of nitrite ion.

The chemiluminescence (CL) method shows 10^6 response selectivity for nitrite ion over nitrate, sulfate, urea, caffeine, ammonium, and chloride. Responses observed for 10% (w/w) aqueous solutions of the above compounds disappeared upon addition of a reagent that reacts with nitrite ion to form a diazonium ion[38] (0.6 M sulfanilamide in 10% hydrochloric acid). This observation suggests that the responses from the 10% solutions were due to nitrite ion contamination of the samples of various potential interferences. The only compound surveyed that generates a signal in the presence of sulfanilamide is S-nitrosoglutathione. It is anticipated that other S-nitrosothiol compounds will respond similarly. The low detection limits, high selectivity, small sample volumes needed, and rapid sampling rate of this novel FIA system make it well suited for application to the rapid analysis of nitrite ion or NO in complex systems such as biological fluids or environmental samples.

3.3. ANALYSIS OF NITROGEN-CONTAINING ORGANIC COMPOUNDS IN LIQUID SAMPLES BY NO CHEMILUMINESCENCE

3.3.1. Background

Nitrogen-containing compounds in petroleum products, wastewater, foods, and the environment play important divergent roles as nutrients and hazardous pollutants. The measurement of these compounds could provide, for instance, a sense of viability of bioremediation or prediction of health risk from exposure to *N*-nitrosamines. Application of NO chemiluminescence to analysis of these complex sample types provides a route to selective detection based on the presence of nitrogen in the compound or compounds of interest.

3.3.2. Analysis of *N*-Nitroso Compounds

3.3.2.1. Principle of Operation

Techniques for the analysis of nitrosamines are based on the cleavage of the N—NO nitroso bond, with subsequent introduction of the resulting NO to a chemiluminescent detector. Two NO chemiluminescence methods have been reported: thermal energy analysis (pyrolysis), and reductve cleavage. The N—NO bond possesses a bond energy on the order of $10-50 \, \text{kcal mol}^{-1}$, compared to bond energies of approximately $80 \, \text{kcal mol}^{-1}$ for C—C, C—H, and C—N bonds. Photolysis or pyrolysis of nitrosamines have been shown to be sufficiently energetic processes to rupture this bond.[11,12,39] Denitrosation of *N*-nitroso compounds was also observed to occur on a rapid timescale at room temperature with a mild reducing agent (10% sodium iodide),[38] and in acidic iodide.[40]

3.3.2.1.1. Thermal Energy Analysis. Pyrolysis and chemiluminescence have been applied to the analysis of nitrosamines in foods, beverages, and biological samples.[11,12,41-44] This technique, termed *thermal energy analysis* by Thermoelectron Corporation, Waltham, Massachusetts, the company that markets the TEA™ analyzer, is based upon the catalytic pyrolysis of the sample to liberate NO from nitrosamines in the sample.

Nitrosamine in a sample is extracted with dichloromethane. The resulting solution is injected into a pyrolyzer maintained at greater than $275\,°C$.[41] The presence of a catalyst reduces the pyrolysis temperature,[43] improving the selectivity of the process; a mixture of WO_3 and $W_{20}O_{58}$ has been recommended as the catalyst of choice.[12,43] Selectivity of the system is high owing to the low pyrolysis temperature, the selectivity of the gas phase ozone chemiluminescent reaction, and the use of an optical filter to allow only wavelengths greater than 600 nm to reach the PMT. Selectivity of the method was evaluated by injection of a large number of compounds. Relatively few compounds yielded responses; those include isopentylnitrite, nitric acid, nitrate ion, nitrite ion, and dimethyl sulfoxide.[41] This technique has a detection limit of 400 fmol of dimethylnitrosamine (a volatile nitrosamine); the calibration curve is linear over 6 decades.[12]

Thermal energy analysis has been coupled with both gas chromatographic (GC) and HPLC techniques.[39,41,42,44,45] The combination of GC and TEA reported in 1975[42] is linear over 5 decades, with sensitivity of 500 ppt of nitrosamine in a 200 µL standard injection. Such large sample injections result in significant peak tailing, but demonstrate the sensitivity of the instrumentation. The authors predicted that previously reported interferences from organic nitrites would be eliminated in this method through decomposition in the hot (200–250 °C) GC injection port. A large response from the solvent is attributed to the presence of impurities in the solvent such as dissolved NO, or organic nitrites that decompose in the injection port. The solvent from includes a negative baseline dip caused by increased collisional quenching of background chemiluminescence by solvent molecules. If this presents a problem, a cold trap (−150 °C) placed between the GC column and the chemiluminescence detector should remove solvent and any other low-volatility pyrolysis products. Since the vapor pressure of NO is in excess of 760 torr at this temperature the concentration of NO would not be affected.[42]

A GC method was used to analyze samples from fried bacon, cured sausage, fish, and cheese products, as well as other complex samples.[39] In order to confirm the presence of nitrosamines, Doerr and Fidler analyzed the samples by GC and TEA, exposed to UV light, then reanalyzed. Nitrosamines present in the samples are photolyzed and are not present in the chromatogram after photolysis. The disappearance of peaks is confirmation of the presence of nitrosamines in the original analysis.

Decomposition of organic nitrates in a hot injection port prevents their analysis by GC. In order to measure these compounds by chromatography with TEA, a HPLC method can be used. The analysis of glycerol trinitrate, pentaerythritol tetranitrate, and isosorbide dinitrate in blood plasma[45] was performed by coupling HPLC and TEA, avoiding the risk of thermal decomposition in a GC injection port. Spiked plasma samples were extracted with ethyl acetate, concentrated, and a 25 µL aliquot was injected into the HPLC. The pyrolyzer of the TEA apparatus was operated at 500 °C, a higher temperature than for nitrosamine analysis, to provide conversion of these organic nitrates to NO. The detection limit was 0.1 ng for each of glycerol trinitrate and pentaerythritol tetranitrate, and 0.2 ng of isosorbide dinitrate.

3.3.2.1.2. Reductive Cleavage of Nitrosamines. Analysis of nitrate and nitrite ion by liquid phase reduction and NO chemiluminescence[15] upon the reaction of NO with O_3 is described earlier in this chapter. By changing the reducing agents and sample preparation techniques, Cox et al. have developed a selective method for total nitrosamine content.[38]

The reductive conditions used in the analysis of nitrate ion and nitrite ion are sufficiently rigorous to reduce nitrosamines to NO. However, in samples containing nitrosamines it is likely that the concentrations of nitrate or nitrite ion will be significantly greater than the concentration of nitrosamine and will mask the presence of nitrosamine. Therefore, the removal of both nitrate and nitrite ion from the sample prior to analysis is required. Two methods were used for nitrate and nitrite ion removal: ion stripping on an anion exchange resin, and chemical reaction of nitrite with sulfanilamide. A Dowex IX-8 anion exchange resin yielded greater than 97% removal of nitrate ion from urine containing between 6.9 and 38.3 ppm of nitrogen present as nitrate. Removal of nitrite ion by the same resin was 84–89% effective for samples containing between 6.8 and 40 ppm nitrogen present as nitrite. More complete removal of nitrite ion was accomplished by addition of acidic sulfanilamide [0.6 M in 10% (v/v) HCl(aq)]. The reaction between nitrite ion and sulfanilamide yields the diazo salt, which does not produce NO in the subsequent reduction; this reaction increased the nitrite ion removal efficiency to greater than 94%.

Once nitrate and nitrite ion levels were reduced to a level at which they did not interfere significantly, 5 mL of the sample was mixed with 13 mL acetic acid, 3.5 mL sulfuric acid, 1 mL of 10% sodium iodide (aq), and about 1 to 2 drops of octanol; the solution was sparged with He until no NO response was observed. Octanol was added to prevent foaming of the sample during sparging.

3.3.2.2. Applications

The reductive method of nitrosamine analysis has been applied to the analysis of total nitrosamine concentration in urine[38] and cutting fluids.[40] Model nitrosamines were N-nitrosodiethanolamine (NDELA) and N-nitrosoatrazine (NNA). Analysis of urine samples spiked with each of these compounds showed recoveries of 95–100% for NDELA and 33–50% for NNA. Authentic urine samples analyzed for total nitrosamines ranged in measured concentrations from 0.84 to 8.51 ppb of nitrogen as nitroso groups. The method is linear over 3 decades and has a detection limit of 3×10^{-8} M N-nitrosamine. The analysis of cutting fluids by this method determined concentrations ranging from less than 0.0001 ppm by weight to 2.1 (ppmw) nitrogen as nitroso groups in commercial cutting fluids. Cutting fluid samples spiked with various concentrations of NDELA yielded recoveries between 80% and 97%.

3.3.3. Analysis of Organonitrogen Compounds

Pyrolysis of nitrogen-containing organic compounds in the presence of oxygen and subsequent formation of NO for detection by chemiluminescence has been reported.[13] The technique, called *pyrochemiluminescence*, is based upon the combustion of nitrogen-containing analyte above 900 °C in the presence of oxygen. The resulting NO is subsequently detected by chemiluminescence upon reaction with ozone.

3.3.3.1. Principle of Operation

Organic compounds containing nitrogen can be combusted at high temperatures in the presence of oxygen to yield NO. The conditions for the combustion are controlled so that it occurs smoothly (i.e., without explosion) and so that the major product is NO and not compounds with nitrogen in other oxidation states. The temperature range of smooth combustion is greater than 700 °C. Temperatures above 900 °C yield the optimum conversion.[13] At these combustion temperatures it was claimed that neither NO$_2$ nor N$_2$O was formed. The combustion products are swept with He or Ar through a drier to remove water to an evacuated chamber for reaction with O$_3$. The resulting chemiluminescence is monitored by a PMT.

3.3.3.2. Applications

Examples of the application of pyrochemiluminescence include the analysis of nitrogen-containing organic compounds in petroleum samples,[46,47] oil shale retort waters,[48,49] and seawater,[50] and as a GC detector.[51] The analysis of nitrogenous compounds in petroleum fractions was performed on both liquid and high-boiling (>550 °C) samples.[46] Liquid samples were injected using a syringe, whereas higher boiling samples were introduced to the instrument in a quartz sample boat. A wide range of nitrogen-containing test compounds was investigated. Responses from solutions containing concentrations of test compounds at 100–1000 ppm were compared to the response produced by 9-methylcarbazole. While recoveries from 92% to 110% were observed for most test compounds, including pyridine, quinoline, indole, and aniline, low recoveries resulted for nitro, nitroso, and azo compounds, including abzobenzene, nitrobenzene, and N-nitrosodiphenylamine. Deviation between this method and the Kjeldahl procedure was within the precision of the methods for injected samples and within ±5% for samples introduced in the quartz boat.[46,47]

Processing of oil shale yields an aqueous fraction into which a variety of compounds from the resulting petroleum-like crude oil are partitioned. These contaminants include ammonia, amines, and N-heterocycles. The ability to analyze total nitrogen and organic nitrogen has been taken as one indication of the efficacy of biodegradation of this waste. Standards prepared from nitrogenous compounds such as potassium nitrate, nicotinic acid, and pyrazole were analyzed by pyrochemiluminescence and the Kjeldahl procedure. In general, the pyrochemiluminescence method yielded quantitative recoveries more often than did the Kjeldahl method.[48] However, statistical analysis of the results from testing real oil shale process waters by both methods showed that Kjeldahl and pyrochemiluminescence analytical results are not significantly different.

The analysis of total dissolved nitrogen in seawater was accomplished using pyrochemiluminescence.[50] Small aliquots of sample (10 µL) were pyrolyzed under pure oxygen at 1110 °C. Nitrate ion and antipyrine were two standard compounds used; comparable sensitivity was observed for both compounds. The analysis of natural seawater samples by pyrochemiluminescence compared well with results obtained by photooxidation.

Coupling a gas chromatograph to a pyrochemiluminescence detector has allowed the analysis of nitrogen-containing analytes eluted from a GC column.[51] The analysis of nitrosamines requires a relatively low temperature (450–600 °C) for pyrolysis, whereas the more stable organic nitrogen-containing compounds are pyrolyzed at 1000 °C. The method allows detection of as little as 12 pg of nitrogen. This method has been used to analyze nitrosamines, pesticides, pharmaceutical products, and nitrogen compounds in food.[51]

3.4. ANALYSIS OF ATMOSPHERIC NITROGEN-CONTAINING COMPOUNDS

3.4.1. Analysis of NO in the Atmosphere

Reactive nitrogen compounds such as NO, NO_2, HNO_3, NO_3, N_2O_5, and peroxyacetyl nitrate (PAN) participate in important atmospheric reactions. Chemiluminescence based on the reaction of NO with ozone has provided a technique that can measure these species either directly, in the case of nitric oxide, or after conversion of the other species to NO. The development of the NO chemiluminescence detector[6] led to direct measurement of nitric oxide in the trophosphere[52,53] and stratosphere[9] at very low concentration levels. One of the nitric oxide measurements in the troposphere was made at the surface of the Pacific Ocean[53] and illustrates the sensitivity of this chemiluminescent technique. Nitric oxide concentrations were all at or below 4 pptv (parts per trillion by volume), some of the lowest atmospheric NO levels ever measured.

The NO chemiluminescence detector can be used to measure the other nitrogen compounds in the atmosphere that can be converted to NO. The concentration of NO_2 in the atmosphere is of interest owing to the roles this compound plays in destruction and formation of ozone and acid rain. Several techniques have been developed for the conversion of NO_2 to NO for detection; each will be discussed below.

3.4.2. Thermal Conversion of NO_x and NH_3 to NO for Detection

The application of the NO chemiluminescence detector to the analysis of reactive nitrogen compounds has been accomplished by the development of techniques for the conversion of the species of interest to NO. The effectiveness of different catalysts and reducing or oxidizing agents (e.g., C, V, Mn, Mo, W, Pt, and Au) for NO_2 reduction and NH_3 oxidation to NO at atmospheric pressure have been investigated.[54] Temperatures for conversion of NO_2 to NO for the various materials ranges from $>98\%$ at 475 °C for V, W, and Mn to 750 °C for 90% conversion on stainless steel 304. The conversion process results in the deactivation of the metal tube over time; consequently the tube must be heated to a higher temperature for quantitative conversion. The use of carbon or carbon/metal composite converters provides a means to quantitatively convert NH_3 and NO_2 to NO. In addition, the NO_2 to NO conversion reaction on the carbon-based converter produces CO, which is swept out and does not foul the conversion tube. A carbon/Mo tube at 475 °C yields 98% conversion of NO_2 to NO for analysis without significant NH_3 conversion, whereas passing the sample stream over a carbon/Cu converter at 400–450 °C converts both NO_2 and NH_3 to NO for determination. The direct analysis of the sample without oxidation or reduction yields the NO concentration, and the concentrations of all three species can be found by difference.

A system based on a hot stainless steel converter allowed for the analysis of NO, $(NO + NO_2)$, and $(NO + NO_2 + NH_3)$ via conversion of NO_2 and NH_3 in the airstream to NO for detection (see Figure 3.6).[17] The lack of NO_2 or NH_3 conversion selectivity on hot stainless steel is controlled by passing the stream over an ammonia trap consisting of 30–40 mesh firebrick treated with acidic $Cr_2O_7^{2-}$ before contact with the steel converter for NO_2 conversion. The sample stream for $NO + NO_2 + NH_3$ analysis passes directly over the converter. Again, the direct measurement of NO allows the calculation of all three species by difference.

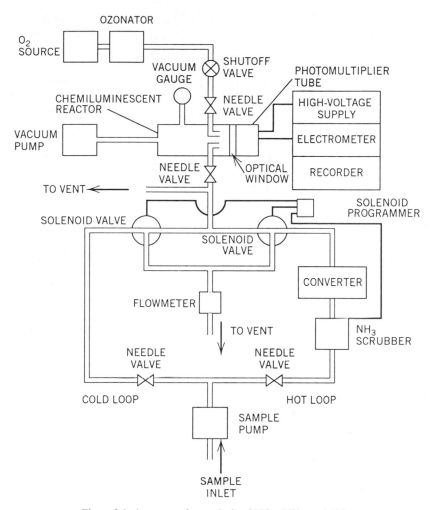

Figure 3.6. Apparatus for analysis of NO_2, NH_3, and NO.

The conversion of nitrogen-containing compounds to NO on Mo and C, in high-temperature converters, was reported by Winer et al.[55] PAN, ethyl nitrate, ethyl nitrite, and n-propyl nitrate all were reported to yield greater than 90% conversion to NO on Mo or C converters; nitroethane showed a 6–7% conversion. While the concentrations of these interfering compounds are low relative to NO_2, unobserved changes in the concentrations of these interferents would lead to high NO_x values and thus erroneously high NO_2 concentrations.

The conversion of HNO$_3$ was observed to occur quantitatively over a Mo surface at 300–350 °C.[56] The analysis of gas phase nitric acid (HNO$_3$) was performed using NO chemiluminescence detection by comparing the response from two hot Mo converters. The gas sample is split: one sample passes over a nylon filter to remove HNO$_3$ before contact with a Mo converter; the second stream flows directly to a second Mo converter. The difference in the signals from the two converters yields a measure of HNO$_3$.

3.4.3. Nonthermal Conversion of NO$_2$ to NO for Detection

3.4.3.1. Reduction of NO$_2$ to NO over FeSO$_4$

The reduction of NO$_2$ in an air sample to NO for detection has been accomplished by passing the sample over a bed of FeSO$_4 \cdot$7H$_2$O.[57–62] Efficiencies of conversion are reported to be between 90% and 95%. The reduction of other nitrogenous pollutants occurs: nitrous acid (HNO$_2$) is converted with 100% efficiency, whereas PAN reduction yields only 10% NO. The relative concentrations of these compounds in air compared to NO$_2$ is negligible in all but the most polluted air or that in smog chamber studies. Of particular interest is the lack of HNO$_3$ conversion on FeSO$_4$.

The conversion process is highly dependent on the humidity of the sample stream. It has been observed that the color of the material lightens when it is exposed to the analytical conditions of temperature and pressure. The dependence on sample humidity for reproducibly quantitative conversion efficiencies of NO$_2$ to NO suggests that the amount of water of hydration present in or on the surface of the FeSO$_4$ plays a role in the conversion process. In order to make the conversion efficiency more nearly constant, the sample stream is humidified.[58]

3.4.3.2. Photolysis of NO$_2$ to NO

Photolysis of NO$_2$ to NO provides more selective conversion than thermolysis in metallic converters.[63] A 300 W high-pressure, short-arc xenon lamp illuminates the sample, generating a steady-state concentration of NO from the NO$_2$ present. The operator intermittently measures NO in the sample by turning off or shuttering the lamp from the sample cell (see Figure 3.7). The photolysis system does not convert N$_2$O$_5$, PAN, chlorine nitrate (ClONO$_2$), and peroxynitric acid (HO$_2$NO$_2$), which yield a NO response through thermal converters. This instrument has been used to measure NO$_x$ (=NO + NO$_2$) in the trophosphere[64] and been compared to the FeSO$_4$ technique for quantitative NO$_x$ analysis. Photolytic conversion of NO$_2$ to NO was found

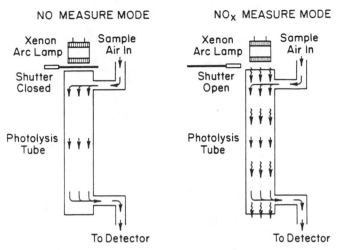

Figure 3.7. NO_2 photolysis system. Sample stream directed to NO chemiluminescence detector immediately after photolysis.

to be free from interferences from PAN or PAN-like compounds[58] and not dependent upon sample humidity.

3.4.4. Catalytic Conversion of NO_y to NO for Detection

The nonselective conversion of reactive nitrogen species in addition to NO_2 (i.e., NO_y) to NO presents a method for the analysis of NO_y that varies with the type of converter system used. When a method for conversion of NO_2 plus all remaining reactive nitrogen species (HNO_3, NO_3, N_2O_5, HONO, HO_2NO_2, PAN, and organic nitrates) is coupled with direct analysis of NO by chemiluminescence, and a selective NO_2-to-NO conversion technique, such as photolysis, the total abundance of atmospheric nitrogen compounds and their distribution can be discerned. For instance, hot Mo tubes convert to NO, HNO_3, HONO, PAN, N_2O_5, and organic nitrates in addition to NO_2.[57] However, memory effects and changes in the zero value have been observed.[59] Another method (see Figure 3.8) has been developed that allows the quantitative conversion of NO_y to NO using carbon monoxide (CO) as a reducing agent on a gold (Au) catalyst [see Reactions (8) and (9)].[8,18]

$$CO + NO_2 \xrightarrow[\Delta]{Au} NO + CO_2 \tag{8}$$

$$CO + HNO_3 \xrightarrow[\Delta]{Au} NO + CO_2 + OH \tag{9}$$

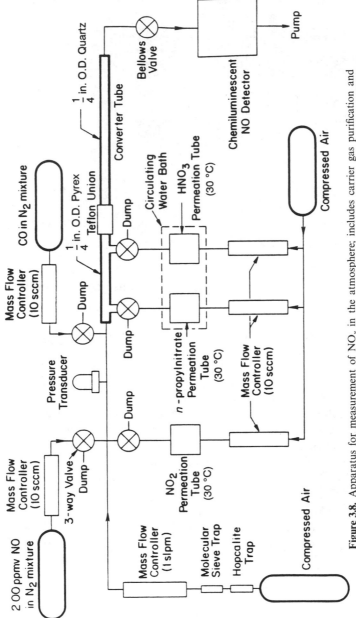

Figure 3.8. Apparatus for measurement of NO_y in the atmosphere; includes carrier gas purification and design of gas phase calibration system: slpm = standard liters per minute; SCCM = standard cubic centimeter per minute.

90

A thorough evaluation of this method was performed by Fahey et al.[19] They used an Au converter at 300 °C and observed more than 90% conversion efficiency for NO_2, HNO_3, N_2O_5, and PAN. This method is now used extensively to measure NO_y species in the troposphere.

3.5. THE REDOX CHEMILUMINESCENCE DETECTOR

3.5.1. Background

The oxidative ability of NO_2 has been demonstrated in the reactions between CO and NO_2 for the generation of NO for detection.[8,18] Oxidation–reduction reactions between NO_2 and other reduced organic and inorganic compounds can also produce NO when performed under the proper conditions with the appropriate catalyst.[20,65,66] The application of NO_2 oxidations to analytical chemistry has been accomplished with the redox chemiluminescence detector (RCD). Catalyzed reactions between NO_2 and analytes of interest generate a surrogate NO signal that is sensitively detected by chemiluminescence. Generation of a surrogate NO signal has been applied to GC,[21,67–69] supercritical fluid chromatography (SFC),[70] and liquid chromatography (LC).[71]

3.5.2. Principle of Operation

Detection by RCD is based upon three steps: separation of the sample chromatographically, catalytic oxidation of the analyte with NO_2, and detection of the resulting NO by ozone chemiluminescence. The operation of the NO chemiluminescence detector has already been discussed in this chapter. While details pertaining to chromatographic separations are outside the scope of this review, the catalytic oxidation of analyte by NO_2 will be the step discussed further here.

The selectivity of the RCD is affected by the catalyst material, catalyst support, catalyst temperature, and makeup gas flow rates. A variety of metals have been investigated as catalysts of the NO_2 oxidation of analytes. Catalytic activity has been exhibited by silver, copper, and rhodium; however, palladium and gold have shown the most activity. Gold and palladium differ significantly in their activity.[69] While alkanes are not oxidized by NO_2 over gold on soda-lime glass at temperatures less than 450 °C, palladium on borosilicate catalyzes this reaction at temperatures as low as 250 °C.[69] A superconducting perovskite ($YBa_2Cu_3O_7$) has been used as a catalyst in the RCD.[72] At temperatures of 200–350 °C, this catalyst afforded sensitive detection of several compounds including alcohols, alkenes, acetaldehyde, methyl ethyl ketone, and ammonia. The detection limit for ethanol was 0.5 ng at a catalyst temperature of 325 °C.[72]

The composition of the support material on catalyst activity has been demonstrated by comparing responses of alkanes to gold on borosilicate glass to gold on soda-lime glass. On gold-coated soda-lime glass, n-alkanes do not rapidly react with NO$_2$ at 400 °C; here the RCD shows high selectivity for detection of oxygenated organics in the presence of alkanes. However, if borosilicate beads coated with gold are used, n-alkanes do react with NO$_2$ at these temperatures.[65]

Both catalyst temperature and makeup gas flow rate affect the sensitivity of the detector. Whereas the sensitivity to alcohols decreases at catalyst temperatures above 275 °C, sensitivity to alkenes increases with increasing temperature. As flow rates increase, the response of benzene decreases to below the detection limit at flow rates higher than 18.8 mL/min. The response for alcohols decreases only slightly. The adjustment of makeup gas flow rate and catalyst temperature adds additional dimensions of selectivity to the system. By adjusting the makeup gas flow rate and catalyst temperature, the RCD was made selective for oxygenates, allowing the additive, methyl tert-butyl ether (MTBE), to be analyzed in the complex gasoline matrix.[72]

3.5.3. Applications

3.5.3.1. Gas Chromatography

The resolving power of GC can provide about 100,000 theoretical plates in a modern capillary column. Even with this extraordinary ability to separate the components of a sample with general detectors, the analysis of a single compound or group of compounds in complex samples can still elude the chromatographer, and so selective detectors are needed. The electron capture detector (ECD) is probably the most extensively used selective detector because of its ability to detector electrophilic halogenated and nitro-containing compounds.

The RCD provides selectivity for groups of compounds that can be readily oxidized, including alcohols, ketones, and acids,[21,67–69] along with compounds not detected sensitively by other methods such as flame ionization detection (FID). These compounds include carbon disulfide, sulfur dioxide, hydrogen, carbon monoxide, and hydrogen peroxide.[67]

A stream of NO$_2$ from a thermostated permeation tube is swept with He and mixed concentrically with the effluent from a GC column (see Figure 3.9). The NO$_2$ permeation tube was used to provide from 5 to 500 ppm NO$_2$ at around 30 mL/min of He.[21,67] So that the resolution provided by the GC column was preserved, a short, low dead-volume catalyst bed was placed

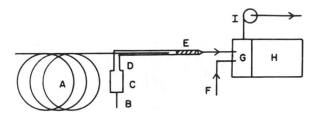

Figure 3.9. Schematic diagram of redox chemiluminescence detector (RCD); A, column from gas chromatograph; B, He carrier; C, NO$_2$ permeation device; D, inlet for dilute NO$_2$ stream; E, catalyst bed for reduction of NO$_2$; F, ozone supply; G, chemiluminescence reaction chamber; H, photomultiplier tube; I, vacuum pump.

directly after the column. In normal operation, the catalyst bed was maintained at 150–400 °C. Labile compounds such as 1-pentanol are converted at catalyst temperatures as low as 200 °C,[20] whereas many other compounds including alcohols, ketones, olefins, and organic acids are detected with the catalyst bed at 400 °C.[67] Alkanes, chlorinated solvents, water, and carbon dioxide do not yield appreciable responses at the normal catalyst operating temperature (400 °C).

The RCD has been applied to the analysis of lemon oil, gasohol, simulated refinery gas, extracts of municipal wastewater, and volatile atmospheric organic compounds. The principal weakness of RCD steams from the variability of catalytic activity, particularly when species are present that poison the gold catalyst surfaces.

3.5.3.2. Supercritical Fluid Chromatography and Liquid Chromatography

Supercritical fluid chromatography (SFC) provides the high resolution of GC along with the ability to analyze less volatile compounds as done by HPLC. Since capillary columns are used in both techniques and flow rates from these columns are comparable, many detectors employed by GC are applicable to SFC. The coupling of SFC with the RCD[70] was achieved by inserting the GC column into the RCD reaction catalyst chamber as previously done for GC. Carbon dioxide (CO$_2$) is used as the supercritical fluid eluent, since it does not reduce NO$_2$ to NO. As the supercritical fluid CO$_2$ leaves the column, the sudden decrease in pressure expands the fluid into the gas phase, carrying eluted analytes directly to the catalyst bed.

Application of the RCD to LC has been accomplished with liquid phase reaction between dilute HNO$_3$ and reduced compounds. The NO produced can be transported to the gas phase for chemiluminescence detection.[71] The reaction occurs between HNO$_3$(aq) and organic or inorganic compounds at

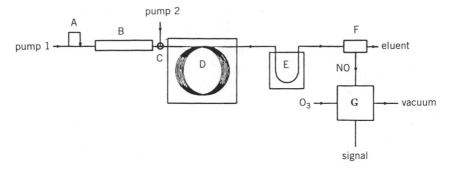

Figure 3.10. Schematic diagram of LC/RCD: A, fixed-loop injection valve; B, LC column; C, mixing tee; D, capillary reaction coil; E, cooling bath; F, pneumatic nebulizer; G, chemiluminescence reaction chamber.

temperatures greater than 150 °C at pressures around 1200 psi (see Figure 3.10). Inorganic compounds that yield a NO response after reaction with HNO$_3$ include Fe(II), iodide, thiocyanate, thiosulfate, nitrite ion, and hydroxylamine.

Organic compounds that reduced HNO$_3$ to NO under these conditions include sugars and ascorbic acid, whereas less reactive compound such as tetrahydrofuran required higher reaction temperatures to effect appreciable conversion of HNO$_3$ to NO. The reaction chamber was constructed of a 30 m × 0.23 mm i.d. uncoated, fused silica capillary column. At a reagent flow rate of 1.0 mL/min, the residence time in the reaction chamber was approximately 60 s. Through change of the flow rate, the residence time of analyte could be adjusted. It was observed that as residence time was increased the response also increased; this indicated that conversions were incomplete, even for the most labile compounds tested. It is important for analytical purposes to have a constant percentage of conversion for each analyte if the conversion is incomplete.

The transport of NO produced in the liquid phase reactions to the gas phase for detection was accomplished with a pneumatic nebulizer. The capillary reaction coil was fitted directly into the junction of a glass tee. Helium was forced concentrically around the capillary, through a constricted orifice at the end of the capillary. The resulting He stream sprayed the eluent into fine droplets, allowing NO(aq) to partition into the gas phase. Transport efficiency for this system with reaction chamber temperature at 225 °C was measured to be 28 ± 1%.

The method was demonstrated by analysis of three saccharides: sucrose, fructose, and ribose. A 20 μL aliquot for standard containing 10 ppm of each of the sugars was separated on a column of Macronex CHO-682 lead(II)-

loaded cation-exchange polymer. The reproducibilities for the resulting NO response peaks were less than 6% RSD. The analysis of saccharides illustrates the utility of this method; the sensitivity is comparable to currently used methods for saccharides. The technique may be particularly useful for compounds lacking chromophores.

REFERENCES

1. R. J. Strutt, *Proc. Phys. Soc., London* **23**, 66–73 (1910).
2. Y. Tanaka and M. Shimazu, *J. Sci. Res. Inst., Tokyo* **43**, 241–251 (1949).
3. J. C. Greaves and D. Garvin, *J. Chem. Phys.* **30**, 348–349 (1959).
4. M. A. A. Clyne, B. A. Thrush, and R. P. Wayne, *J. Chem. Soc., Faraday Trans.* **60**, 359–370 (1964).
5. P. N. Clough and B. A. Thrush, *J. Chem. Soc., Faraday Trans.* **63**, 915–925 (1967).
6. A. Fontijn, A.J. Sabadell, and R. J. Ronco, *Anal. Chem.* **42**, 575–579 (1970).
7. H. S. Johnston and H. J. Crosby, *J. Chem. Phys.* **22**, 689–692 (1954)
8. M. J. Bollinger, Chemiluminescent measurements of the oxides of nitrogen in the clean troposphere and atmospheric chemistry implications. Doctoral Dissertation, University of Colorado, Boulder (1982).
9. B. A. Ridley and L. C. Howlett, *Rev. Sci. Instrum.* **45**, 742–746 (1974).
10. R. L. Shearer and R. E. Sievers, in *Chemiluminescence and Photochemical Reaction Detection in Chromatography* (J. W. Birks, ed.), pp. 72–97. VCH Publishers, New York (1989).
11. D. H. Fine, F. Rufeh, and B. Gunther, *Anal. Lett.* **6**, 731–733 (1973).
12. D. H. Fine, F. Rufeh, and D. Lieb, *Nature (London)* **247**, 309–310 (1974).
13. R. E. Parks and R. L. Marietta, U.S. Pat. 4,018,562 (1977).
14. O. C. Zafiriou, M. McFarland, and R. H. Bromund, *Science* **207**, 637–639 (1980).
15. R. D. Cox, *Anal. Chem.* **52**, 332–335 (1980).
16. O. C. Zafiriou and M. McFarland, *Anal. Chem.* **52**, 1662–1667 (1980).
17. J. E. Sigsby, F. M. Black, T. A. Bellar, and D. L. Klosterman, *Environ. Sci. Technol.* **7**, 51–54 (1973).
18. M. J. Bollinger, R. E. Sievers, D. W. Fahey, and F. C. Fehsenfeld, *Anal. Chem.* **55**, 1980–1986 (1983).
19. D. W. Fahey, C. S. Eubank, G. Hübler, and F. C. Fehsenfeld, *J. Atmos. Chem.* **3**, 435–468 (1985).
20. S. A. Nyarady and R. E. Sievers, *J. Am. Chem. Soc.* **107**, 3726–3727 (1985).
21. S. A. Nyarady, Reduction-oxidation chemiluminescence detection. Doctoral Dissertation, University of Colorado, Boulder (1985).
22. M. A. Franson, *Standard Methods for the Examination of Water and Wastewater*, 14th ed. Am. Public Health Assoc., Washington, DC, 1975.

23. Z. Iskandarani and D. J. Pietrzyk, *Anal. Chem.* **54**, 2601–2603 (1982).

24. H. J. Cortes, *J. Chromatogr.* **234**, 517–520 (1982).

25. R. G. Gerritse, *J. Chromatogr.* **171**, 527–529 (1979).

26. D. T. Gjerde, J. S. Fritz, and G. Schmuckler, *J. Chromatogr.* **186**, 509–519 (1979).

27. D. T. Gjerde, G. Schmuckler, and J. S. Fritz, *J. Chromatogr.* **187**, 35–45 (1980).

28. P. E. Jackson, P. R. Haddad, and S. Dilli, *J. Chromatogr.* **295**, 471–478 (1984).

29. S. H. Kok, K. A. Buckle, and M. Wootton, *J. Chromatogr.* **260**, 189–192 (1983).

30. A. Mangia and M. T. Lugari, *Anal. Chim. Acta* **159**, 349–354 (1984).

31. C. Garside, *Mar. Chem.* **11**, 159–167 (1982).

32. K. Yoshizumi, A. Kazuyuki, T. Matsuoka, and S. Asakura, *Anal. Chem.* **57**, 737–740 (1985).

33. A. R. Thornton, J. Pfab, and R. C. Massey, *Analyst* (*London*) **114**, 747–748 (1989).

34. T. Aoki, *Biomed. Chromatogr.* **4**, 128–130 (1990).

35. A. J. Dunham, R. M. Barkley, and R. E. Sievers, *Anal. Chem.* Accepted for publication (1994).

36. J. M. Reijn, W. E. Van Der Linden, and H. Poppe, *Anal. Chim. Acta* **126**, 1–13 (1981).

37. T. Sunden, A. Cedergren, and D. D. Siemer, *Anal. Chem.* **56**, 1085–1089 (1984).

38. R. D. Cox, C. W. Frank, L. D. Nikolaisen, and R. E. Caputo, *Anal. Chem.* **54**, 253–256 (1982).

39. R. C. Doerr and W. Fiddler, *J. Chromatogr.* **140**, 284–287 (1977).

40. R. D. Cox and C. W. Frank, *Anal. Chem.* **54**, 557–559 (1982).

41. D. H. Fine, F. Rufeh, D. Lieb, and D. P. Rounbehler, *Anal. Chem.* **47**, 1188–1191 (1975).

42. D. H. Fine and D. P. Rounbehler, *J. Chromatogr.* **109**, 271–279 (1975).

43. D. H. Fine, D. Lieb, and F. Rufeh, *J. Chromatogr.* **107**, 351–357 (1975).

44. D. H. Fine, G. S. Edwards, I. S. Krull, and M. H. Wolf, in *Nitrogenous Air Pollutants: Chemical and Biological Implications*, (D. Grosjean, ed.), pp. 55–65. Ann Arbor Sci. Publ., Ann Arbor, MI,, 1979.

45. W. Yu and U. Goff, *Anal. Chem.* **55**, 29–32 (1983).

46. H. V. Drushel, *Anal. Chem.* **49**, 932–939 (1977).

47. J. Y. Shay and P. W. Woodward, *Am. Lab.* (*Fairfield, Conn.*) **18**, 114–123 (1986).

48. B. M. Jones and C. G. Daughton, *Anal. Chem.* **57**, 2320–2325 (1985).

49. C. G. Daughton, B. M. Jones, and R. H. Sakaji, *Anal. Chem.* **57**, 2326–2333 (1985).

50. T. W. Walsh, *Mar. Chem.* **26**, 295–311 (1989).

51. L. O. Courthaudon and E. M. Fujinari, *LC-GC* **9**, 732–734 (1991).

52. D. H. Stedman, E. E. Daby, F. Stuhl, and H. Niki, *J. Air Pollut. Control Assoc.* **22**, 260–263 (1972).

53. M. McFarland, D. Kley, J. W. Drummond, A. L. Schmeltekopf, and R. H. Winkler, *Geophys. Res Lett.* **6**, 605–608 (1979).

54. L. P. Breitenback and M. Shelef, *J. Air Pollut. Control Assoc.* **23**, 128–131 (1973).

55. A. M. Winer, J. W. Peters, J. P. Smith and J. N. Pitts, Jr., *Environ. Sci. Technol.* **8**, 1118–1121 (1974).

56. D. W. Joseph and C. W. Spicer, *Anal. Chem.* **50**, 1400–1403 (1978).

57. R. R. Dickerson, A. C. Delany, and A. F. Wartburg, *Rev. Sci. Instrum.* **55**, 1995–1998 (1984).

58. F. C. Fehsenfeld, R. R. Dickerson, G. Hübler, W. T. Luke, L. J. Nunnermacker, E. J. Williams, J. M. Roberts, J. G. Calvert, C. M. Curran, A. C. Delany, C. S. Eubank, D. W. Fahey, A. Fried, B. W. Gandrud, A. O. Langford, P. C. Murphy, R. B. Norton, K. E. Pickering, and B. A. Ridley, *J. Geophys. Res.* **92**, 14710–14722 (1987).

59. R. R. Dickerson, *J. Geophys. Res.* **90**, 10,739–10,743 (1985).

60. R. R. Dickerson, *Atoms. Environ.* **18**, 2585–2593 (1984).

61. R. J. Kelly and D. H. Stedman, *Geophys. Res. Lett.* **6**, 375–378 (1979).

62. J. A. Ritter, D. H. Stedman, and T. J. Kelly, in *Nitrogenous Air Pollutants: Chemical and Biological Implications* (D. Grosjean, ed.), pp. 325–343. Ann Arbor Sci. Publ., Ann Arbor, MI, 1979.

63. D. Kley and M. McFarland, *Atmos. Technol.* **12**, 63–69 (1980).

64. D. Kley, J. W. Drummond, M. McFarland, S. C. Liu, *J. Geophys. Res.* **86**, 3153–3161 (1981).

65. N. Pourreza, S. A. Montzka, R. M. Barkley, R. E. Sievers, and R. S. Hutte, *J. Chromatogr.* **399**, 165–172 (1987).

66. R. E. Sievers and S. A. Nyarady, U.S. Pat. 4,717,675 (1988).

67. S. A. Nyarady, R. M. Barkley, and R. E. Sievers, *Anal. Chem.* **57**, 2074–2079 (1985).

68. R. E. Sievers, S. A. Nyarady, S. Barak, S. S. Banning, and R. S. Hutte, *Chromatography*, pp. 673–677 (1985).

69. R. E. Sievers, S. A. Nyarady, R. L. Shearer, J. J. DeAngelis, and R. M. Barkley, *J. Chromatogr.* **349**, 395–403 (1985).

70. W. T. Foreman, R. E. Sievers, and B. W. Wenclawiak, *Fresenius' Z. Anal. Chim.* **330**, 231–234 (1988).

71. R. E. Sievers, R. M. Barkley, and J. J. DeAngelis, *J. Chromatogr.* **441**, 125–134 (1988).

72. E. A. McNamara, S. A. Montzka, R. M. Barkley, and R. E. Sievers, *J. Chromatogr.* **452**, 75–83 (1988).

CHAPTER

4

THE OXYGEN-SENSITIVE DETECTOR O-FID

ULRICH K. GÖKELER

Siemens AG, 76181 Karlsruhe, Germany*

4.1. INTRODUCTION

Chromatography is the widest used analytical technique that enables not only the separation of components of interest but also the selective quantification of individual components. Because of the complexity of mixtures frequently analyzed and the limitation in regard to resolution, despite the use of high-resolution capillary columns and multicolumn systems to reduce the complexity of the separation, specific detectors are used to gain selectivity and simplify the separation system. Over the years many different detectors have been invented and developed that have sensitivities toward specific atoms or certain functional groups in molecules.

Specifically for the selective detection of oxygen-containing molecules, methods such as nuclear magnetic resonance (NMR),[1,2] infrared (IR) spectrometry,[3,4] microwave-induced helium plasma,[5–7] gas chromatography–atomic emission detection (GC–AED),[8] GC–mass spectrometry (GC–MS),[9], high-performance liquid chromatography, (HPLC)[10,11] or other techniques[12] were reported. But because of the complexity, price and demands on the operator, these methods are not widely accepted.

Use of a single high-resolution capillary column is insufficient owing to a lack of selectivity when it is confronted with the high number of components present in typical samples, and it is therefore limited to a few oxygenates.[13,14]

By using multidimensional GC, the technique of coupled columns, any known oxygenate can be separated and detected individually even in a complex matrix, even fully automated.[15–24] However, multidimensional GC

*Current address: ES Industries, Division for Siemens Chromatographs, 701 South Route 73, Berlin, New Jersey 08009.

Selective Detectors, edited by Robert E. Sievers. Chemical Analysis Series, Vol. 131.
ISBN 0-471-01343-9 © 1995 John Wiley & Sons, Inc.

is unsuccessful if unknwon oxygenated hydrocarbons have to be identified and separated without the help of the aforementioned more complicated hyphenated techniques to identify them.

Over the past 10 years or so, triggered by the demand for a simple analytical device to detect oxygenated hydrocarbons in fuel, a principle that was

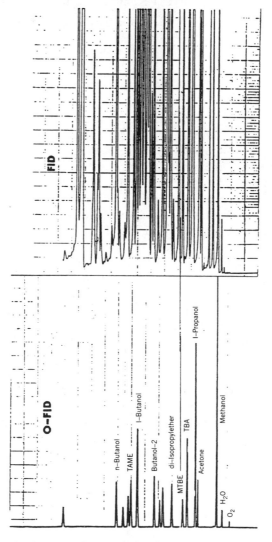

Chromatogram 4.1. Simultaneous detection with O-FID and FID: gasoline spiked with oxygenates. (From Gökeler,[44] courtesy of the publisher.)

investigated and used more than 50 years ago[25] was technically refined, resulting in an oxygen-specific detector called the O-FID. As the acronym indicates, this detector is based on the commonly used flame ionization detector (FID). However, with the changes incorporated, it is sensitive to any type of molecule containing at least one oxygen atom, totally neglecting the presence of any molecule containing no oxygen within the operating parameter. By means of O-FID, the separation task is reduced to a single-column, single-detector configuration that simplifies the detection of known and unknown oxygenated hydrocarbons in complex matrixes tremendously; therefore O-FID can be routinely used in analytical laboratories without the need for specialists to operate it (Chromatogram 4.1).

4.2. PRINCIPLES OF OPERATION

In addition to the basic analytical features such as the gas supply, injection system, separation column, and FID, the oxygen-sensitive detector consists of the following two distinctive parts (Figure 4.1):

- The reaction or cracking reactor, which disintegrates the molecules and generates carbon monoxide (CO)
- The methanizer, which converts CO into methane (CH_4), detected by means of the FID

Any oxygen-containing molecule entering the cracking reactor will be disintegrated and generates carbon monoxide according to

$$C_nH_mO_x \xrightarrow{+\Delta H} xCO + (n-x)C + (m/2)H_2 \tag{1}$$

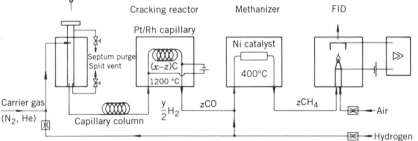

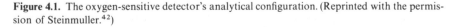

Figure 4.1. The oxygen-sensitive detector's analytical configuration. (Reprinted with the permission of Steinmuller.[42])

Because there is a significant excess of carbon generated, carbon accumulates and remains inside the cracking capillary. In addition, because of the high temperature, the Boudouard equilibrium [Equation (2)][26] and the water equilibrium [Equation (3)][27] have shifted in favor of CO, so that the remaining H_2O and CO_2 generation produced by

$$C + CO_2 \overset{+\Delta H}{\rightleftharpoons} 2CO \qquad (2)$$

and

$$C + H_2O \overset{+\Delta H}{\rightleftharpoons} CO + H_2 \qquad (3)$$

can be completely neglected.

Naturally there is the possibility of methane generation according to

$$C + 2H_2 \overset{+\Delta H}{\rightleftharpoons} CH_4 \qquad (4)$$

But amounts are negligible as well and do not vary under typical conditions ($> 1200\,K$ and 1 bar) if there is significantly more hydrogen than oxygen in the carrier gas.[28]

Any hydrocarbons not containing oxygen that enter the cracking reactor will be disintegrated according to

$$C_nH_m \overset{+\Delta H}{\longrightarrow} nC + (m/2)H_2 \qquad (5)$$

Therefore, excessive carbon is deposited, excessive hydrogen flows unnoticed through the FID, and only oxygen is converted into CO by using one carbon for each oxygen. The O-FID is typically used for hydrocarbon matrixes, which provide an excess of carbon with each injection, and thus fulfills the requirements for Equations (2) and (3). The carbon deposit inside the reactor provides carbon also for non-hydrocarbons such as oxygen in air, elemental oxygen (O_2), and water (H_2O), where carbon and water are converted into carbon monoxide according to Equation (3) and O_2 (or oxygen in air) according to

$$C + O_2 \overset{-\Delta H}{\rightleftharpoons} CO_2 \qquad (6)$$

and then, according to Equation (2), to CO.

If oxygenates are detected that are not present in a hydrocarbon matrix and therefore need excess carbon, the carbon deposit has to be regenerated continuously; otherwise it will be consumed and the requirements of Equations (2) and (6) no longer be met.

The generated carbon monoxide is converted into methane using the methanizer connected downstream right after the cracking reactor according to

$$CO + 3H_2 \xrightarrow{+\Delta H + \text{catalyst}} CH_4 + H_2O \qquad (7)$$

which subsequently is detected by the FID.

Therefore, each oxygen atom present as part of a molecule or in the form of elemental oxygen is converted and detected as methane and the analysis is simplified to the separation of the oxygenated hydrocarbons. This makes the separation and quantification extremely simple (Chromatogram 4.1).

4.3. DESIGN

4.3.1. The Cracking Reactor

The design principle of the cracking reactor goes back to the determination of the total amount of oxygen which was reported by Schütze,[25] Unterzaucher,[29] and others.[30-35]

In order to determine, among other elements, the total oxygen content in hydrocarbon matrixes, the sample was disintegrated at 1400 K in a glass tube under a nitrogen atmosphere and the oxygen was converted to CO in the presence of coal. Quantification was done using wet chemistry owing to the unavailability of chromatography or other automated analytical methods.

Whereas this procedure was used exclusively for the total amount of oxygen, Kaiser[36] applied the same principle to the determination of individual oxygenated hydrocarbons, first using GC to separate the individual components, with subsequent conversion to CO/CH_4 to get the selectivity. Kaiser's method, first mentioned in a report in 1975[36] and further elaborated upon in 1977,[37] employed a heated micro-packed stainless steel tube filled with carbon and a methanizer to detect oxygenated hydrocarbons selectively and simultaneously with O-FID and FID. The problems involved with oxygen contamination and the lack of a need for such a selective detector did not make it a success at that time.

In a 1982 publication by Schneider et al.,[38] an O-FID was described that had all the elements of an easy-to-use oxygen-specific detector using capillary columns for the separation task. In the following years some additional reports were published that described either modifications or experience with the O-FID.[39-44]

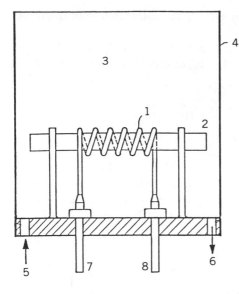

Figure 4.2. Cracking reactor design: 1, coiled reactor capillary; 2, support carrier; 3, MgO heat-damping material; 4, reactor housing; 5, inlet for purge gas; 6, purge gas vent; 7, inlet from capillary column; 8, exit to methanizer.

The cracking reactor (Figure 4.2) consists of a heated platinum–rhodium (Pt/Rh 90/10) alloy capillary tube with typical dimensions of 1 mm o.d., 0.5 mm i.d., and some 400–500 mm length. The Pt/Rh alloy is used owing to its physical strength at temperatures between 1250 and 1650 K for an acceptably long life, its capability of being formed mechanically into a tube with the flexibility to be coiled, its being gas tight, as well as the ease with which it can be heated and conduct thermal energy. Any catalytic effect originating from the Pt/Rh alloy is of no importance for this process because after the first few injections the surface will be completely covered with carbon.

Other materials used in the past were glass tubes[29] or steel tubes filled with carbon.[36]

The reactor tube is either coiled around a support material[38,46] or just immobilized at its ends.[43,45,47] The reactor tube, which is insulated at the gas chromatograph, is heated by applying current on its ends using a high-current/low-voltage power supply. Depending on the exact dimensions, typical currents of 9–11 A and 4–6 V [35–65 VA] are applied that heat the reactor tube to 1250–1650 K.

The temperature distribution along the reactor tube is of importance because it influences the cracking capability of the reactor tube, cracking reproducibility, and lifetime. In order to provide a reasonably similar temperature profile along the reaction tube, the tube is equidistantly coiled around an aluminum oxide support[38,46] and/or the volume around the reaction tube is filled with a heat-insulating material such as magnesium oxide power.[45,47]

The required reaction temperature depends on the cracking capacity desired and is determined experimentally. The stability of the cracking temperature down to single degrees is obviously not of paramount importance as long as it is exceeding the threshold cracking capacity. However, two types of design have been described: one uses a temperature-controlled heating system with temperature sensor and temperature display;[41] the other is a simpler design that displays the supplied heating current/voltage only but does not give an explicit temperature reading.[45]

Owing to the high temperature, the platinum alloy is destroyed very quickly if not held under an inert environment without the presence of oxygen (rapid mass loss of platinum in the presence of oxygen and temperatures > 1200 K).[47] Therefore, either the reactor tube is in a housing and costantly purged with inert gas or it must be enclosed in a vacuum tube. Because the purged housing design is rugged and simple, it is the design used. Nitrogen or argon is used as purge gas. Owing to its high thermal conductivity, which vents too much energy away from the reactor capillary, helium cannot be used.

Nevertheless, the Pt/Rh reactor tube has a limited lifetime because at the high temperature used it becomes brittle and tiny leaks are generated owing to the recrystallization process taking place. The influence of sublimation of the material under the conditions used seems to be secondary because of lack of oxygen present. The lifetime depends directly on the reaction temperature applied. Excluding premature losses of the reaction tube due to "poisoning" or plugging due to improper usage, operating times of more than 3500 h are typical and more than 10,500 h have been reported.

The latest improvements on the design of the cracking reactor indicates that the reaction tube was modified in such a way that the cracking temperature required is substantially lower. Because it is done at a lower temperature, the cracking of the molecules and the conversion into CO is based on catalytic effects rather than on thermal cracking. This has distinct advantage that the reactor lifetime increases significantly, but, on the other hand, it is easier to poison the reactors catalyst. Detailed information and operating parameters and experience are not available at this time.

4.3.2. The Methanizer

The effluent from the cracking reactor flows through a deactivated fused silica capillary to the methanizer, and the carbon monoxide is converted into Methane according Equation to (7) in the presence of hydrogen, an appropriate catalyst, and elevated temperature. The catalyst typically consists of activated nickel or ruthenium[38,48] on a carrier material such as Al_2O_3.

Earlier designs incorporated the methanizer elegantly as a part of the FID either by inserting a tube filled with the catalyst into the FID[38,41] or by filling the quartz glass FID jet with the catalytic material.[42,47] However, over the years it became apparent that a serious problem could arise with this design when the FID tube was not correctly filled with the catalyst. Because the maximum FID temperature of 400 °C was not evenly distributed along the FID, "cold" spots could occur along the catalyst, resulting in conversion loss and peak tailing.[47] Secondly, and much more important, the total catalyst mass was small and catalyst poisoning frequently occurred due to the presence of sulfur in the sample.[49]

Owing to the residence time of carbon monoxide in the catalytic material inside the FID jet, for example, less than 0.3 s, any loss of methanizer efficiency has an immediate impact on the conversion rate.

This resulted in a design where the methanizer has its own independently heated housing or is integrated into the cracking reactor. Both solutions permit the use of more catalytic mass, a more uniform temperature distribution, and a higher methanizer temperature, which all contributes to a longer methanizer lifetime.

Additionally, there is the advantage of using the FID at a temperature lower than 400 °C, which improves detector performance and lifetime.

4.3.3.   Gas Supply

As already mentioned, it is extremely important to use carrier gas that is free of oxygen (O_2), moisture (H_2O), carbon monoxide (CO), and carbon dioxide (CO_2). This leads to the requirement that the supply gases (carrier gas and hydrogen) be cleaned by filters that remove any contamination containing O_2 and any other oxygen-containing contaminants such as those just mentioned. Naturally, the prevention of any oxygen diffusion from the environment into the carrier gas is also required. Such diffusion happens frequently owing to leaky injector septums or because the column connector fittings are not oxygen-diffusion tight, especially when temperature programming is applied. Even diffusion opposing the carrier gas flow, e.g., from a leaky split connection, can flow into the column and has been found to be a source of contamination and interferences with the measurement. It has been shown that in the trace-level analysis of oxygenates even the diffusion of oxygen through graphite seals can reduce the sensitivity significantly when low-ppm levels of oxygenates are being analyzed.

Another source of diffusion is the pressure regulators used to maintain a constant gas supply within the GC. Only diffusion-free metal pressure regulators are recommended.

4.4. PERFORMANCE

4.4.1. Analytical Parameters

Carrier gas: Helium or nitrogen, $\geqslant 99.999\%$
(O_2, CO, CO_2, H_2O filtered)
(Sum $O_2 < 0.1$ ppmv)
with 0.5–3.0% v/v hydrogen, $> 99.999\%$
(Sum O_2, 0.1 ppmv)
(O_2, CO, CO_2, H_2O filtered)
Flow: 0.5–1.5 mL/min

Purge gas: Nitrogen or argon, $\geqslant 99.995\%$
Flow: 50–100 mL/min

Methanizer: Hydrogen $\geqslant 99.999\%$
(O_2, CO, CO_2, H_2O filtered)
Flow: 20–30 mL/min

Columns: Bonded capillary columns
0.20–0.32 mm i.d.

Reactor: Platinum/rhodium 90%/10%
400–500 mm, 1 mm o.d., 0.5 mm i.d.
1250–1650 K
35–54 VA

Methanizer: SS 316 tube
Activated nickel or ruthenium prepared on aluminum oxide
620–900 K

Basic ionization: $< 5 \times 10^{-12}$ A
Sensitivity: 1.5×10^{-11} g oxygen
Linearity: $> 5 \times 10^3 \ (\pm 1.5\%)$
Dynamic range: 10^5
Selectivity: 10^6:1 (O:C)
Reactor lifetime: 3500–10,000 operating hours ($\geqslant 1250$ K)

4.4.2. Dynamic Range

The dynamic range of the O-FID system depends directly on the cracking capacity of the reactor. The cracking capacity, along with the sensitivity, is directly dependent on the cracking temperature, on the type of molecule to be disintegrated, and on the residence time (i.e., the carrier gas flow rate) inside the hot reaction tube under the given design parameter. This means that the dynamic range can be increased and the sensitivity improved as well by

Table 4.1. Cracking Capacity at 1600 K[a]

Component	Cracking Capacity (μg)	References
Toluene	40–50	38, 43
n-Heptane	10–20	38, 43, 46
1,3-Butadiene	2–4	
Methane	4.7	46

[a] Methyl silicone, 0.32 mm i.d., 60 °C isotherm; 31 cm/s helium.

increasing the reaction tube cracking temperature or reducing the carrier gas flow. However, an important drawback is that the lifetime of the reactor tube is directly proportional to the cracking temperature and a fixed carrier gas flow rate must be maintained in order to have optimum separation efficiency.

Therefore, in practice, to accommodate the required dynamic range and the cracking capacity, the cracking temperature is adjusted in such a way that the desired sensitivity is just reached but not exceeded. Any further increase would enhance the cracking capacity and therfore the sensitivity but would reduce the lifetime of the reactor tube significantly.

The cracking capacity, in general, is lower for aliphatic hydrocarbons than for aromatic hydrocarbons and decreases even more for short-chain olefins (Table 4.1).

4.4.3. Linearity

Linearity measurements for specific oxygenated hydrocarbon have shown that correlations are typically between 0.99 and 0.9999 over a range of 10^3–10^5, depending on the reaction tube temperature.[41,43] Ranges reported are 1–3200 ng 2-propanol (0.22–832 ng O_2)[46] and 3–1600 ng tert-butyl alcohol (TBA),[38] respectively.

However, users in routine laboratories, where a compromise between cracking capacity, sensitivity, and lifetime of the reactor tube has to be made, report typical acceptable linearities in the range of 2–5×10^3.

4.4.4. Sensitivity

The sensitivity, i.e., the signal-to-noise ratio (S/N), of the O-FID is dictated by two factors: the reactor tube temperature, which dictates the cracking capacity (hence the sensitivity), and the oxygen residue in the carrier gas, which dictates the background noise. Since oxygen is the only molecule detected, the sensitivity is proportional to the relative amount of oxygen mass in the molecule. This means that the higher is the molecular weight,

Table 4.2. Experimentally Determined Sensitivity for Various Oxygenated Hydrocarbons

Component		Molecular Weight (g/mol)	O Contents (%)	Sensitivity (ppm)	Oxygen (ppm)
Methanol	CH_3OH	32.04	49.9	~12	~6
Ethanol	C_2H_5OH	46.07	34.7	~17	~6
1-Propanol	C_3H_7OH	60.10	26.6	~24	~6
1-Butanol	C_4H_9OH	74.12	21.6	~28	~6
1-Pentanol	$C_5H_{11}OH$	88.15	18.1	~33	~6
MTBE[a]	$C_5H_{12}O$	88.15	18.1	~33	~6
ETBE[b]	$C_6H_{14}O$	102.18	15.7	~38	~6
1-Hexanol	$C_6H_{13}OH$	104.18	15.7	~39	~6
1-Heptanol	$C_7H_{15}OH$	116.20	13.8	~43	~6
1-Octanol	$C_8H_{17}OH$	130.23	12.3	~49	~6

[a]MTBE = methyl tert-butyl ether.
[b]ETBE = ethyl tert-butyl ether.

the lower the molar sensitivity of the molecule. Table 4.2 shows the dependence of sensitivity on the percentage of O in the molecule, which results in an oxygen sensitivity of approximately 6 ppmv (0.15 ng O_2). One way to improve the sensitivity of the O-FID is expanding the single column with a multicolumn separation system. By applying "heart-cuts," i.e., venting the hydrocarbons before they enter the cracking reactor, the injection volume can be increased significantly under the given conditions and the sensitivity of the oxygenated hydrocarbons can be increased by more than a factor of 10 without overloading the cracking capacity of the reactor. However, with such a configuration only known oxygenates can be monitored, and a diffusion-free column switching system is required as well.[49]

4.4.5. Selectivity

The selectivity ratio of the O-FID between oxygen and hydrocarbons under the given design are dependent on the cracking capacity, hence on the reaction tube temperature as well as on the oxygen residue in the carrier gas (background noise). Typically, the ratio is 10^6:1 (O:C),[41] but by increasing the cracking capacity it can be as high as 10^7:1 (O:C).[38]

4.4.6. Response Factors

According to Equations (1) and (7), for each oxygen in a molecule, one methane is generated. Therefore, if the conversion rate is independent of the individual molecule, the response should be the same as long as the number

Table 4.3. Detector Response for Various Oxygenated Hydrocarbons, Normalized to Methanol

Component	Molecular Weight (g/mol)	O Contents (%)	Relative O Response
Methanol	32.04	49.91	1.00
Ethanol	46.07	34.71	0.97
Isopropanol	60.10	26.61	0.99
Isobutanol	74.12	21.57	0.97
2-Butanol	88.15	21.57	0.97
TBA	74.12	21.57	0.98
2-Pentanol	88.15	18.14	0.94
MTBE	88.15	18.14	0.98
TAME[a]	102.18	15.65	0.97
Acetone	58.08	27.53	0.97
Methyl ethyl ketone	72.11	22.18	0.99
Methyl acetate	74.08	43.17	0.99

Source: Reprinted with permission from Steinmuller.[43]

[a]TAME = *tert*-amyl methyl ether.

of oxygen atoms is the same. The experimental results normalized to methanol presented in Table 4.3 show that the difference in response for various oxygenated hydrocarbons is small (RSD 1.5%).

4.4.7. Reproducibility

The reproducibility of the detector was investigated with regard to concentration and retention time using a standard O-FID system with an autosampler,

Table 4.4. Concentration and Retention Time Reproducibility Over 20 h (59 Consecutive Injections)[a]

	Ethanol	MTBE	TAME	TAA[b]
Concentration				
Average area counts	29360	14277	16650	12171
Range	1141	647	989	590
RSD (%)	1.14	1.06	1.38	1.14
Retention times (min)	5.68	9.56	12.83	16.15
Range (min)	0.01	0.01	0.02	0.04
RSD (%)	0.45	0.34	0.69	0.98

[a]For the analytical parameters see Chromatogram 4.3.

[b]TAA = 2-methyl-2-butanol (*tert*-amyl alcohol).

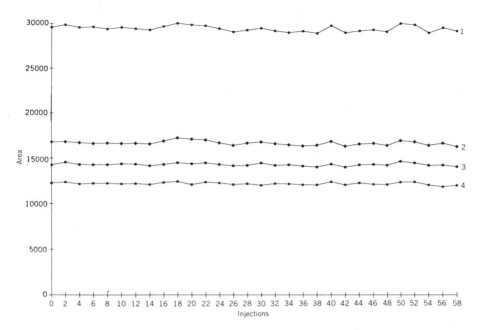

Figure 4.3. Reproducibility graph over a 20 h period from 59 consecutive injections (only every second data point is shown): 1, ethanol, RSD = 1.14%; 2, TAME, RSD = 1.38%; 3, MTBE, RSD = 1.06%; 4, TAA, RSD = 1.14%.

split injection, and a single methyl silicone column. The data shown in Table 4.4 and Figure 4.3 were generated using the suggested analytical parameters from the draft EPA method[51] for O-FID. The data were accumulated over a 20 h period from 59 consecutive injections from the same spiked gasoline sample. Similar performance data have been published previously.[38,43,46,47,49]

4.4.8. The Cracking Temperature

As outlined earlier, the cracking capacity is influenced directly by the reaction tube temperature under constant parameters. This was impressively demonstrated by Schneider et al.[38] in a series of chromatograms where the reaction tube temperature was increased step by step from 873 to 1473 K and where the response of the hydrocarbon mixture decreased with increasing temperature till none of the hydrocarbons were any longer visible.

For a certain required sensitivity, hence cracking capacity, the reaction tube temperature is adjusted in such a way that under given analytical parameters all hydrocarbons are disintegrated (Chromatogram 4.2). Any additional cracking capacity (i.e., higher reaction tube temperature) leads

```
─START IF

1 ┌─  2.15
2 ┤   2.47

    3.16




STOP

RUN #    65

AREA%
   RT           AREA  TYPE     AR/HT      AREA%
  2.15         28934   PB     0.029      60.675
  2.47         18753   PB     0.059      39.325

TOTAL AREA=        47687
MUL FACTOR=  1.0000E+00
```

Chromatogram 4.2. "Blank" chromatogram of gasoline. Only air (1) and water (2) are visible; all other hydrocarbons have completely disintegrated.

Table 4.5. Cracking Temperature Influence on the Detector Response[a]

	Reaction Tube Temperature			
	1450 K	1550 K	1650 K	ST (%)
Methanol	1	1.012	1.037	1.6
Isopropanol	0.323	0.330	0.324	1.3
TBA	0.473	0.465	0.469	0.9
MTBE	0.784	0.778	0.795	1.1

[a]Normalized to methanol at 1450 K.

only to shorter lifetime but does not generate any analytical benefits. Table 4.5 demonstrates that there is insignificant influence on the oxygen response when the reaction tube temperature varies. Therefore, precise monitoring of the reaction tube temperature is not absolutely necessary as long as the cracking threshold is exceeded.

4.4.9.   The Cracking Reactor and Methanizer Poisoning

As is the situation whenever catalysts are used, under certain circumstances they can be poisoned reversibly or irreversibly. Different kinds of poisoning are encountered that can be triggered by oxygen, hydrogen, high-molecular-weight hydrocarbons, sulfur, phosphorus, lead, chlorine, or silica.

In order to prevent excessive oxygen from coming in contact with the Pt/Rh reaction tube at the required high temperature, the outside is purged with nitrogen as already described. Despite the presence of oxygen as traces in the carrier gas (i.e., when the analytical system is used with oxygen contamination of the carrier gas) and in the measuring components, it has not been shown to have any influence on the Pt/Rh after the carbon deposit has been generated.

Adding hydrogen to the carrier gas as described above promotes the possibility of faster recrystallization. Initial tests have not conclusively proven that the lifetime of the reaction capillary is prolonged by not using hydrogen in the carrier gas. High-molecular-weight components such as those present in crude oil are sometimes not disintegrated entirely inside the reaction capillary owing to insufficient residence time and the residue can be carried into the methanizer, poisoning it reversibly. Sulfur, lead, phosphorus, and chlorine sample constituents, as well as silica from methyl silicone column bleed, can prevent reproducible conversion inside the reactor if present in significant quantities. Silica is also known to contribute to the process of the Platinum becoming brittle. In addition, especially when a methanizer with a small catalytic mass is used, sulfur has been shown to poison the methanizer irreversibly. Sulfur quantities typically present in the 300–500 ppmv range in gasoline have not been shown to create any problems even when sulfur-containing samples are injected repetitively over a longer period of time (several months).[42] However, the methanizer should be checked frequently with a known standard when mixtures containing sulfur are used.

4.5.   APPLICATIONS

Presently, there are two major fields where the oxygen-specific detector is used:

- Automotive applications such as petroleum products, especially gasoline, and automotive emissions
- Natural products

4.5.1. Automotive Applications

The majority of applications for which O-FIDs are used today have to do with automotive matters. The reason is that the O-FID's development was triggered by the simplicity with which it selectively detects a group of com-

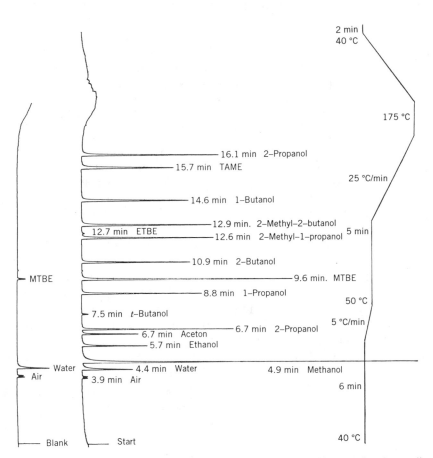

Chromatogram 4.3. Blank gasoline spiked with possible gasoline additives, analyzed according to the proposed EPA method.[50] *Analytical parameters*: Methyl silicone capillary column: 60 m long; 0.25 mm i.d., 1 μm film thickness. Injection volume 1 μL. Split 1:100. Reactor temperature 1340 K. Temperature program as indicated. Component concentration 0.5–2.0%.

ponents in a rather complex matrix; it was subsequently supported by the method proposed by regulatory authorities[51,52] to monitor quality standards in gasoline for a wide variety of possible components (Chromatogram 4.3).

Oxygenated hydrocarbons are presently added to gasoline because they reduce exhaust emissions especially during winter in certain geographic areas, and improve engine performance. Because there may be a wide variety of oxygenated hydrocarbons added and the permissible additive concentration levels are rather tightly prescribed, the O-FID is the ideal analytical device to monitor any additive.

Chromatogram 4.4 shows four different gasolines of various brands from the same geographic area.

Other applications involve the determination of oxygenated hydrocarbons in naphtha (Chromatogram 4.5), kerosene extract (Chromatogram 4.6), and crude oil (Chromatogram 4.7). All these samples typically have some levels of sulfur the presence of which has not been shown to have any negative effects on the O-FID.

Further reported applications using the selectivity of the O-FID are the determination of furfurol at the 10 ppm level,[47] which is used to extract aromatic hydrocarbons in lubrication oil, or the monitoring of alcohol, ketones, ethers, and aldehydes during the gasoline-manufacturing process[46] where, owing to the low concentration of these components, an extraction/ enrichment procedure is used before injection.

Oxygenated hydrocarbons in engine exhaust emission can exceed the capabilities of the O-FID because the concentration levels of oxygenated hydrocarbons are at the low-ppm/high-ppb level, which is the limit of O-FID capabilities even when coupled columns are used.[53] Therefore, only a combination with on-line sample enrichment permits the successful use of the O-FID; this is presently under investigation in several laboratories.

A sample enrichment step prior to analysis was used in order to determine oxygen-containing volatile aromatic hydrocarbons (PAH) in brown coal emission condensate of residential furnaces in the eastern part of Germany.[54] By using O-FID's selectivity, O-PAH's fingerprint was identified and confirmed by using GC/MS.

4.5.2. Natural Products

Whereas there are only a few oxygenated hydrocarbons as opposed to a high number of non-oxygenated hydrocarbons present in petroleum products, the situation with natural products is the reversed. Here there are only a few non-oxygenated hydrocarbons and plenty of oxygenated hydrocarbons. According to Sandra et al.,[54] typically most flavor and fragrance active components are oxygenated hydrocarbons. Therefore, the aim with flavor

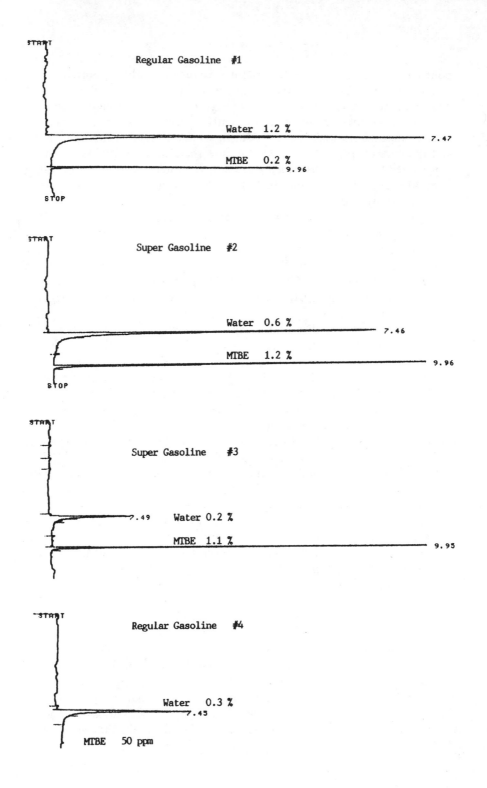

START
Regular Gasoline #1

Water 1.2 %
7.47
MTBE 0.2 %
9.96
STOP

START
Super Gasoline #2

Water 0.6 %
7.46
MTBE 1.2 %
9.96
STOP

START
Super Gasoline #3

7.49 Water 0.2 %
MTBE 1.1 %
9.95

START
Regular Gasoline #4

Water 0.3 %
7.45
MTBE 50 ppm

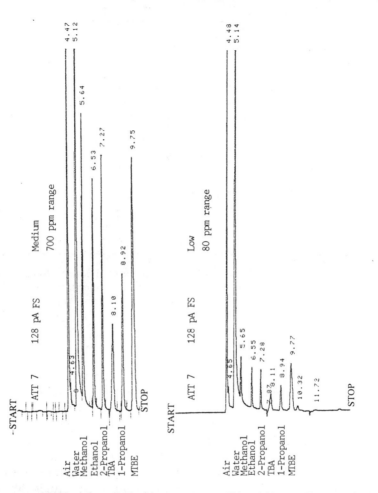

Chromatogram 4.5. Naphtha spiked with oxygenates at 700 and 80 ppm. The "negative" peak between 2-propanol and TBA is due to elevated levels of oxygen in the carrier gas and the presence of Hydrocarbons at high concentration levels. *Analytical parameters*: Methyl silicone capillary column: 40 m long; 0.32 mm i.d. 5 μm film thickness. Temperature program: 90 °C for 3 min, 10 °C per min to 180 °C. Injection volume 1 μL. Split 36:1. Nitrogen carrier gas, velocity 33 cm/s. Reactor temperature ∼1300 K. (FS = Full scale; ATT = Attenuation.)

Chromatogram 4.4. MTBE and water contents in gasoline from various commercial filling stations, September 1991. *Analytical parameters*: Methyl silicone capillary column: 100 m long; 0.25 mm i.d. 0.5 μm film thickness. 60 °C isothermal. Injection volume 1 μL; Split 40:1. Helium carrier gas, velocity 25 cm/s. Reactor temperature ∼1300 K.

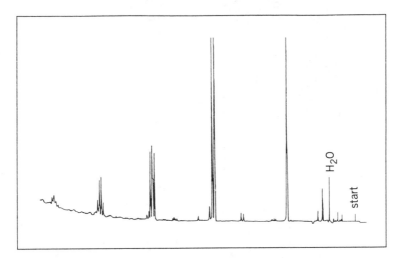

Chromatogram 4.6. Oxygenated hydrocarbons in kerosene extract. *Analytical parameters:* Carbowax 20 M capillary column: 30 in. long; 0.32 mm i.d. Temperature program: 50 °C for 3 min, 5 °C min to 250 °C. Injection volume 1 μL. Split 30:1. Reactor temperature ~1450 K.

and fragrance components is the determination of non-oxygenated hydrocarbons by eliminating them and comparing the chromatogram with a complete scan generated by an FID to determine the hydrocarbons.

Lakszner et al.[56] demonstrated the simultaneous use of the O-FID and FID as quality control tools to fingerprint chamomile oil from various geographic areas and adulterated oils as well. Also, by comparing French and English lavender as well as natural and synthetic bergamot on lemon oil[57] in addition to orange oil,[41] the ease with which these tools can be used for application and evaluation is demonstrated.

Chromatogram 4.8 shows the simultaneous detection of a limonene ($C_{10}H_{16}$) extract with the FID and O-FID. It is clearly apparent that the first group of components that is not visible at all on the O-FID must consist of hydrocarbons only. The main peak in this group is limonene. The majority of the oxygenated hydrocarbons is located in the middle group, while only a few are in the last group.

Applying the O-FID to perfumes proved not to be very successful because, along with the solvent used, all other compounds are typically oxygenated hydrocarbons and virtually no non-oxygenated hydrocarbons exist as long as the perfume is of a certain acceptable quality (Chromatogram 4.9).

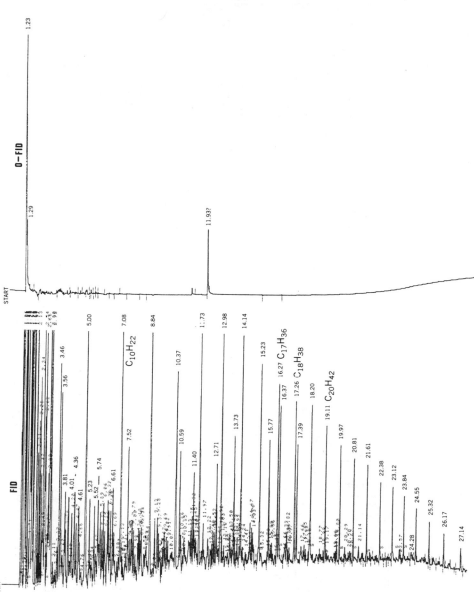

Chromatogram 4.7. Single individual oxygenated hydrocarbon of significant size found in Patoka Sweet Crude Oil simultaneously by FID and O-FID. *Analytical parameters*: Methyl silicon column: 30 m long; 0.25 mm i.d.; 0.25 film thickness. Temperature program: 60 °C for 4 min; 12 °C/min to 300 °C for 30 min. Helium carrier gas, velocity 40 cm/s. Split 1:130. Injection volume 0.4 µL. Reactor temperature ∼1480 K.

Chromatogram 4.8. Limonene extract analyzed by FID and O-FID. *Analytical parameters:* Methyl silicone capillary column: 50 m long; 0.25 mm i.d.; 1 μm film thickness. Split 1:75. Injection volume 1 μL. Temperature program unknown.

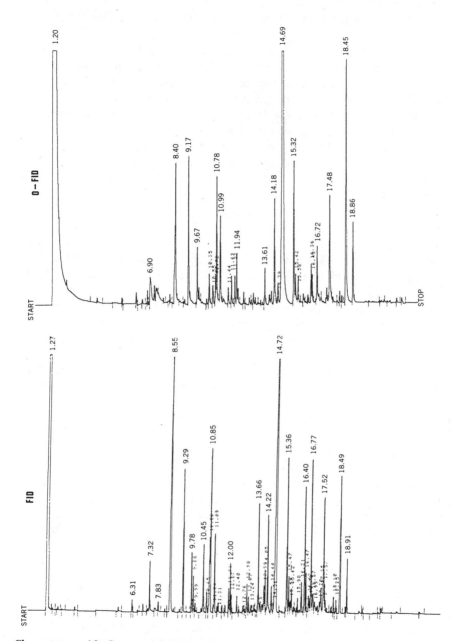

Chromatogram 4.9. Commercial, high-quality perfume analyzed by FID and O-FID. For the analytical parameters, see Chromatogram 4.7.

4.6.   SOME REMARKS AS TO O-FID OPERATION

Unlike more straightforward detectors such as the FID or the thermal conductivity detector (TCD), the O-FID demands more from the operator in regard to understanding the function and the requirement of the detector.

As already outlined, in order to get a stable and low-noise baseline and to prevent any reduction in the cracking capacity due to a constant background conversion, any oxygen-containing impurity in the carrier gas must be avoided. Oxygen not only contributes to the depletion of the carbon deposit and to unreproducible conversion, it also is a source of "unexplained negative" peaks, as mentioned in some reports.[46,47] "Negative" peaks are generated when hydrocarbons of significant concentration displace or reduce the presence of the contaminated carrier gas. Therefore, when this hydrocarbon peak travels through the reactor, there is a lack of contaminated carrier gas; then, because hydrocarbons do not generate a response, the baseline dips owing to the missing oxygen contamination from the carrier gas at the FID. This can be seen nicely in Chromatogram 4.5 between 2-propanol and TBA. The purity of the carrier gas can easily be guaranteed by use of appropriate filters in the supply gas lines. However, any oxygen diffusion from the environment into the separation system is difficult to find, especially because even very small leaks will have a significant influence on the baseline (0.3 ppmv of oxygen increases the basic ionization by 1 pA).

Because these leaks are so small, they cannot be detected by the traditional "bubble"-type or thermal conductivity leak detection systems. It has been proven very helpful to use CO_2 gas, which is blown at suspicious spots. The CO_2 diffuses into the carrier gas and is detected downstream after conversion into methane at the FID.[49]

Before initially starting up the cold reactor or after a prolonged time when it has been sitting idle without being purged, it is recommended to purge the reactor housing for 1–2 days at low temperature; it is better to leave the purge on all the time. Owing to the hygroscopic nature of the magnesium oxide inside the reactor housing, it will be saturated with moisture and conduct the heat so well that the required reaction tube temperature cannot be achieved and there will be a significant temperature gradient along the reaction tube, resulting in incomplete and irreproducible conversion.

Establishment of the proper reaction tube temperature is achieved by a few successive injections of a blank hydrocarbon mixture and a step-by-step increase of the reactor temperature after selection of the analytical parameters such as flow rate, split, injection amount, and starting temperature.

The threshold temperature changes over time owing to the alteration of the reaction tube and has to be checked frequently using a blank hydrocarbon mixture. When the cracking capacity is not sufficient, hydrocarbon "break-

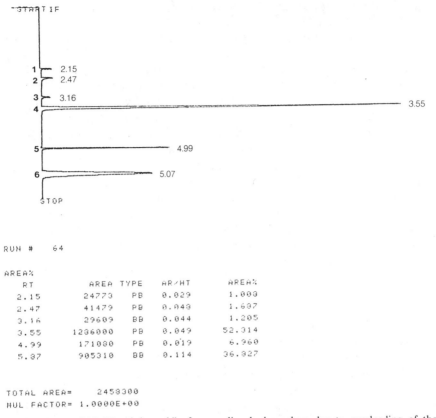

RUN #   64

```
AREA%
    RT          AREA   TYPE    AR/HT      AREA%
   2.15         24773    PB    0.029      1.003
   2.47         41479    PB    0.049      1.687
   3.16         29609    BB    0.044      1.205
   3.55       1286000    PB    0.049     52.314
   4.99        171080    PB    0.019      6.960
   5.07        905310    BB    0.114     36.827
```

```
TOTAL AREA=    2458300
MUL FACTOR= 1.0000E+00
```

Chromatogram 4.10. "Breakthrough" of a gasoline hydrocarbon due to overloading of the reactor cracking capacity: 1, air; 2, water; 3, ethanol, 375 ppm; 4, acetone, 1.15%; 5, "breakthrough" hydrocarbon; 6, ETBE, 0.5%.

through" peaks appear (Chromatogram 4.10). The breakthrough peaks are easily recognized because the much sharper peak forms represent only the hydrocarbon peak tips that exceed the cracking capacity.

If an O-FID system is used that does not have a continuous carbon supply to ensure a permanent carbon deposit inside the reaction tube, it may happen that the deposit vanishes after several days when the system is not in use, owing to oxygen residue in the carrier gas. This may also happen if samples are used that have a lack of excessive carbon. It is then required to frequently inject 2–3 µL of hydrocarbons so as to build up the carbon deposit.

On the other hand, when hydrocarbon samples are used that deposit a high amount of carbon in the reaction tube and when many samples are analyzed, it may happen that the reaction tube is entirely filled with carbon and blocked. This can be observed when the retention time suddenly increases from analysis to analysis because the flow restriction inside the reaction tube increases and slows down the carrier gas. Carbon reduction by means of injection air or water is then required in order to save the reaction tube.

Since the hydrocarbons are not generating a signal on the FID, it is typically not required to purge all the high boiling hydrocarbons from the separation column after each run as long as the remaining hydrocarbons do not reduce the efficiency of the separation columns. However, when the hydrocarbon residue from the column is cleaned out, it is advisable not to heat up the separation column to its maximum permissible temperature, but rather only to approximately 200 °C (e.g., for methyl silicone phases). Owing to the increased column bleed of a methyl silicone column at higher temperatures, silica formed inside the reactor is permanently deposited there; moreover, when a glycol-type separation phase is used, the baseline rises as temperature increases due to the oxygen contents of the column bleed, as can be seen in Chromatogram 4.6.

4.7. SUMMARY

An oxygen-specific detector (O-FID) has been described that permits the selective detection of any oxygenated hydrocarbons by means of a single capillary column. The O-FID significantly reduces the chromatographic separation task of detecting oxygen-containing molecules. The sensitivity level is in the low-ppm range, with a "practical" linear range over more than 3 decades and a reproducibility of better than 1.5% RSD. When oxygen-free supply gases are ensured and oxygen diffusion into the separation system is prevented, the detector is quantitatively stable over several months and has a reactor lifetime of at least 2500 operating hours.

REFERENCES

1. H. C. Steinmetzer and W. Baumeister, *GIT Fachz. Lab.* **24**, 95–100 (1980).
2. G. W. Renzoni, E. G. Shankland, J. A. Gaines, and J. B. Callis, *Anal. Chem.* **57**, 2864–2867 (1985).
3. R. A. Cochran and D. E. Hillman, *J. Chromatogr.* **287**(1), 197–201 (1984).
4. J. L. Wong and B. Jaselski, *Analyst (London)* **107**, 1282–1285 (1982).
5. C. Bradley and J. W. Carnahan, *Anal. Chem.* **60**, 858–863 (1988).
6. K. J. Slatkavitz, P. C. Uden, and R. M. Barnes, *J. Chromatogr.* **355**(1), 117–126 (1986).

7. S. Pedersen-Bjergaard and T. Greibrokk, *J. High Resolut. Chromatogr.* **15**(10), 677–681 (1992).

8. B. D. Quimby, V. Giarrocco, J. J. Sullivan, and K. A. McClearly, *J. High Resolut. Chromatogr.* **15**(11), 705–709 (1992).

9. R. Orlando and B. Munson, *Anal. Chem.* **58**(13), 2788–2791 (1986).

10. R. E. Pauls and R. W. McCoy, *J. Chromatogr. Sci.* **19**, 558–561 (1981).

11. R. E. Pauls, *J. Chromatogr. Sci.* **23**, 437–441 (1985).

12. E. Pella and B. Colombo, *Anal. Chem.* **44**(9), 1563–1571 (1972).

13. V. K. Agarwal, *Analyst (London)* **113**, 907–909 (1988).

14. J. A. Price and K. J. Saunders, *Analyst (London)* **7**, 829–834 (1984).

15. J. M. Levy, *J. Chromatogr. Sci.* **22**(4), 149–152 (1982).

16. *Oxygenates in Gasoline*, ASTM D-4815. ASTM. Philadelphia, 1992.

17. L. Green and Z. Naizhong, Tech. Paper No. 107. Hewlett Packard. Presented at the 36th Pittsburgh Analytical Conference (27) 1985.

18. S. W. S. McCreadie, D. F. K. Swan, G. M. Ogle, R. Pintus, *Proc. Int. Symp. Capillary Chromatogr., 6th, 1985*, pp. 456–465 (1985).

19. A. F. Lockwood and B. D. Caddock, *Chromatographia* **17**(2), 65–69 (1983).

20. N. G. Johansen, *HRC & CC, J. High Resolut. Chromatogr. Chromatogr. Commun.* **7**, 487–489 (1984).

21. Appl. Note 20. Determination of Oxygenated Components in Gasoline. Siemens AG, Karlsruhe, Germany, 1988.

22. L. A. Luke and J. R. Ray, *Analyst (London)* **109**, 989–992 (1984).

23. Z. Naizhong and L. E. Green, *Proc. Int. Symp. Capillary Chromatogr., 6th, 1985*, pp. 466–475 (1985).

24. J. Sevcik, *HRC & CC, J. High Resolut. Chromatogr. Chromatogr. Commun.* **3**, 166–168 (1980).

25. M. Schütze, *Z. Anal. Chem.* **118**, 245 (1939).

26. F. E. Regnier and J. C. Huang, *J. Chromatogr. Sci.* **8**, 267–271 (1970).

27. G. J. Thompson, H. J. Colemean, R. J. Hopkins, C. C. Ward, and H. T. Hall, *Z. Anal. Chem.* **179**, 127 (1961).

28. R. Hahn, K. J. Huttinger, and W. Vohmann, *Chem.-Ing.-Tech.* **51**(5), 501–502 (1979).

29. J. Unterzaucher, *Ber. Dtsch. Chem. Ges.* **73**, 391 (1940).

30. G. Kainz and E. Wachenberger, *Fresenius' Z. Anal. Chem.* **222**, 278 (1966).

31. G. Pippel and S. Romer, *Mikrochim. Acta*, pp. 1039–1046 (1966).

32. I. Klesment, *Mikrochim. Acta*, pp. 1237–1245 (1969).

33. E. Pella, *Mikrochim. Acta*, pp. 13–22 (1968).

34. A. Gotz and H. Bober, *Fresenius' Z. Anal. Chem.* **187**, 92 (1961).

35. L. V. Kutznetsova, E. N. Scotlyarova, and S. L. Dopytschin, *J. Anal. Chem. USSR (Engl. Transl.)* **20**, 836 (1965).

36. R. E. Kaiser, *O-FID*, Short Report to Siemens. Inst. Chromatogr., Bed Durkheim, Germany, 1975.

37. R. E. Kaiser and R. Rieder, *Proc. Int. Clean Air Congr.*, *4th, 1977*, pp. 451–454 (1977).

38. W. Schneider, J. C. Frohne, and H. Bruderreck, *J. Chromatogr.* **245**, 71–83 (1982).

39. J. C. Frohne, *Erdol Kohle* **37**, No. 3 (1984).

40. D. Steinmüller and D. Leonhard, *LaborPraxis* **9**, 11 (1987).

41. G. R. Verga, A. Sironi, W. Schneider, and J. C. Frohne, *HRC & CC, J. High Resolut. Chromatogr. Chromatogr. Commun.* **11** (248), 899–908 (1988).

42. D. Steinmuller, *GIT Fachz. Lab.*, pp. 456–461 (1989).

43. D. Steinmuller, *Am. Lab.* (*Fairfield, Conn.*) **21**, 120–125 (1989).

44. U. K. Gökeler, *Am. Lab.* (*Fairfield, Conn.*) pp. 28D–28F (1992).

45. *Siemens O-FID*, Operating Manual, Chapter 8, Siemens AG, Karlsruhe, Germany, 1990.

46. W. Hermanns, Graduation Theses, Fachhochschule Aachen, Germany (1987).

47. C. Zielinsky, Graduation Theses, Fachhochschule Hamburg, Germany (1989).

48. M. B. Colket, D. W. Naegeli, F. L. Dryer, and I. Glassman, *Environ. Sci. Technol.* **8**(1), 43–46 (1974).

49. F. P. DiSanzo, *J. Chromatogr. Sci.* **28**, 73–75 (1990).

50. U. K. Gökeler and F. Muller, *Proc. Int. Symp. Capillary Chromatogr.*, *8th, 1987*, Vol. 1, pp. 518 (1987).

51. U.S. EPA, test for the Determination of Oxygenates in Gasoline. Ann Arbor, Michisan.

52. DIN 51413 TEIL 7: Gaschromatographische Analyse (O-FID). Beuth Verlag, Berlin, 1990.

53. H. Müller, Oxygenates in Automotive Emission. RGC202 Appl. S412. Siemens AG, Karlsruhe, Germany, 1993.

54. T. Knobloch and W. Engewald, *High Resolution Chromatog.* **16**, 239–242 (1993).

55. P. Sandra, S. Talat, G. Redant, M. Godfroot, M. Verstappe, and M. Verzele, *HRC & CC, J. High Resolut. Chromatogr. Chromatogr. Commun.* **3**, 107 (1980).

56. K. Lakszner, L. Szepesy, I. Torok, and E. Csapo-Barthos, *Chromatographia* **30**(1–2), 47–50 (1990).

57. K. Lakszner and L. Szepesy, *Chromatographia* **26**, 91–96 (1988).

CHAPTER

5

METAL SELECTIVITY BY ION MOBILITY DISCRIMINATION IN A FLAME IONIZATION DETECTOR

MICHAEL M. GALLAGHER and HERBERT H. HILL, JR.

Department of Chemistry, Washington State University, Pullman, Washington 99164-4630

Metal selectivity in a flame ionization detector was first discovered in 1972 during investigations of flame photometric responses of metals.[1] This detector, called a hydrogen atmosphere flame ionization detector (HAFID), differed from a normal FID by employing an atmosphere of hydrogen doped with part per million (ppm) quantities of SiH_4 and with the collecting electrode positioned some 7–8 cm above the flame.[2] In this configuration the HAFID detected organometallic compounds of aluminum, iron, tin, lead, and manganese at picogram (pg) and below quantities. Response selectivity for organometallic compounds vs. pure hydrocarbons was at least 4 orders of magnitude. Applications of the HAFID for the determination of organometallic compounds in complex matrixes have included the detection of organolead and organomanganese antiknock additives in gasoline[3] and the determination of organotin pesticides in marine paint and on apple orchard leaves.[4]

5.1. OVERVIEW OF DETECTOR RESPONSE

In one parametric study, a correlation between optimal electrode height and optimal SiH_4 doping concentration was found.[5] Optimal response was found when the SiH_4 doping concentration was held at 4 ppm in a H_2 gas flow of 3 L/min with the electrode positioned 7 cm above the flame jet. Optimal flame gas flows were found to be 120 mL/min for oxygen, 165 mL/min for air, and 40 mL/min for nitrogen, the packed column carrier gas. During the electrode height study, low electrode height caused reduction of the collected

Selective Detectors, edited by Robert E. Sievers. Chemical Analysis Series, Vol. 131.
ISBN 0-471-01343-9 © 1995 John Wiley & Sons, Inc.

ion current with an offset electrode location. This observation was explained by the idea that the response ions for metals were of lower ionic mobilities and less effectively collected at the offset electrode than were the background and hydrocarbon ions. In a separate parameter investigation after capillary chromatography, optimal gas flows were found to be 2.3 L/min of hydrogen doped with 10 ppm of SiH_4, a combined flow of 260 mL/min for oxygen and nitrogen with a O_2/N_2 ratio of 1.7.[6] Recently, a design modification of the HAFID has enabled the optimal flow of the hydrogen to be reduced by an order of magnitude from about 3 L/min to about 300 mL/min.[7] This reduction in hydrogen flow is of practical significance and was achieved by incorporating a third gas supply line to the detector, providing an inert gas flow to sweep out combustion products in a sheathed flow configuration.

Response ions of the HAFID have been identified by mass spectrometry (MS).[8] For hydrocarbons and for the flame background, response ions were found to be $(H_2O)_nH^+$ ($n = 1-6$); for organometallic compounds, they were identified as $(H_2O)_mSiOH^+$ ($m = 0-3$) and $(H_2O)_x(SiO_2)_yH^+$ ($x = 0-4$; $y = 1-4$). Surprisingly, in no case were metal, metal oxides, or any metal-containing ion species observed when the electrode was positioned farther than 1 cm from the flame jet. The shift in response ions from the hydronium ion clusters observed in the background to a series of protonated silica clusters observed for the metal-containing compounds was believed to be responsible for the organometallic selective response.

These observations led to the thesis that differences in ionic mobility between the background flame ions and the apparently less mobile response ions for organometallics contribute to the selectivity of response. That is, the more mobile background ions (and hydrocarbon response ions) diffuse

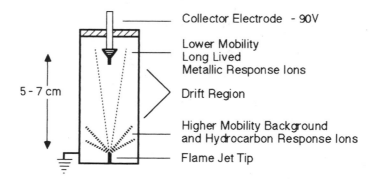

Figure 5.1. Collapsing cone mechanism for organometallic response in the HAFID without biasing electrode.

rapidly to the grounded detector walls whereas the less mobile organometallic response ions remain in the mass flow above the flame to be collect at the high-electrode position. This proposed mechanism can be thought of as a wide cone of higher mobility background flame ions that, in the presence of flame products of organometallic compounds, transfer their charge to ions of lower mobility, resulting in a narrow cone of ions that survive the drift to the collector electrode. Figure 5.1 provides a pictorial representation of this "collapsing cone" mechanism for HAFID response.

To test this collapsing cone concept, a detector designed to discriminate between mass flow and electrical field interactions was constructed. In this study, response selectivity for organometallic compounds over hydrocarbon compounds was found to be further enhanced by more than an order of magnitude by the addition of a biasing electrode to the walls of the HAFID.[9] The studies reported in this chapter were designed, through computer modeling and parametric investigation, to describe in more detail this novel approach to response selectivity in a flame ionization detector.

5.2. EXPERIMENTAL CONSIDERATIONS

5.2.1. Instrumentation

Except for modifications described for individual experiments, the HAFID used for these experiments is shown in Figure 5.2 and described in detail by Gallagher and Hill.[9] The detector gas flows were the following: a total hydrogen flow of 335 mL/min with a SiH_4 doping concentration of 35 ppm was directed to the inner ring of holes in the baffle plate; in addition to the column effluent, the jet tip flow contained 50 mL/min of O_2 and 115 mL/min of N_2 prepared by dilution of pure oxygen with air; also a sheath flow of 650 mL/min of N_2 was vented through the outer ring of holes in the baffle plate. The collecting electrode height was set to 70 mm above the jet tip, and the biasing electrode height was positioned as described in the experiments.

All chromatography was performed on a Hewlett Packard 5880A gas chromatograph (Avondale, Pennsylvania) equipped with a 30 m × 0.25 mm × 0.25 μm d_f DB-1 column from J & W Scientific (Folsum, California). The carrier gas was hydrogen at a flow rate of 0.9 mL/min. The injection port temperature was held at 275 °C, with the detector temperature for both FID and HAFID held at 300 °C. The external walls of the HAFID were heated to 175 °C by heat tape to prevent the condensation of water. For each run the chromatographic oven was operated with a temperature program of 155 °C initially, held for 1.5 min, then ramped at 30 °C/min to 225 °C, where it was held for 1 min.

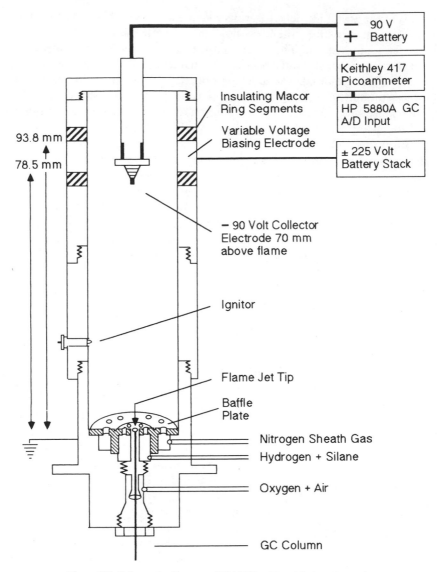

Figure 5.2. Schematic diagram of HAFID with a biasing electrode.

5.2.2. Reagents

Tetra-*n*-butyltin and biscyclopentadienyliron (ferrocene) were purchased from Aldrich Chemical Co. (Milwaukee, Wisconsin) and used without further purification. Tetradecane and hexadecane were purchased from Sigma Chemical

Co. (St. Louis, Missouri). Standards were diluted in hexane obtained from J. T. Baker (Phillipsburg, New Jersey) as necessary. Test mixture concentrations were adjusted to produce peaks of roughly equal size, and daily dilutions were made from refrigerated stock solutions. The mass of each element on column was as follows: iron, injected as ferrocene, 2 pg; tin, injected as tetrabutyl tin, 570 p; and carbon, injected as tetradecane, 124 ng. Gases used were purchased from Liquid Air (Spokane, Washington), with the exception of the 92 ppm SiH_4 in H_2, which was purchased from Airco Specialty Gases (Santa Cruz, California).

5.2.3. Biasing Electrode Experiments

The relationship among positions of and potentials on the biasing and collecting electrodes with respect to metal selectivity were investigated by performing the following experiments:

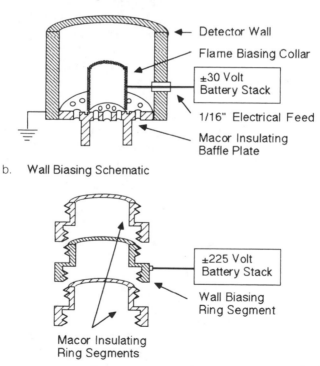

Figure 5.3. Schematic diagrams of two biasing electrode configurations: (a) collar-biasing electrode; (b) wall-biasing electrode.

The first set of laboratory experiments placed the bias voltage directly on the sheathed flow collar. As shown in Figure 5.3a the sheathed flow collar was located around the flame jet and was designed to separate the sheath flow gas from the flame gases. For electrical isolation of this internal collar, a baffle plate was machined out of Macor ceramic and a stainless steel spacer ring segment of the wall had a $\frac{1}{16}$ in. hole bored through it to permit a Teflon insulated wire to be connected to the internal collar. Voltages on the collar were varied from -30 to $+30$ V, and the response of the test compounds were measured.

In the second set of experiments the internal collar was removed from the detector and the bias voltage was placed on a movable segment of the detector was as shown in Figure 5.3b. In these studies the position of the bias electrode was varied from 15 to 94 mm above the jet tip and the voltage was varied from -225 to $+225$ V at each position by a series of batteries.

5.2.4.   Computer Modeling Experiments

Computer modeling of the electric fields generated by the application of bias voltages was accomplished with the MacSimion™ (Montech Pty., Ltd., Clayto, Victoria, Australia) ion trajectory simulation program performed on an Apple Macintosh personal computer (Apple Computer, Cupertino, California). Although the program was designed to calculate and display ion trajectories in a vacuum, the program module that calculates the electric fields and displays equipotential contours imposed by quadrupoles, ion lenses, etc. is valid at atmospheric pressures. With the electric fields accurately represented, the electrostatic forces experienced by the response ions can be described. Ion trajectories under atmospheric conditions are orthogonal to equipotential contours lines of the electric fields.

5.3.   RESULTS AND DISCUSSION

5.3.1.   Collar Biasing Experiments

Figure 5.4 depicts the experimental response from the application of voltage to the internal collar normally used to separate the active SiH_4-doped H_2 atmosphere from the outer sheath gas of N_2. When greater than 30 V of either positive or negative polarity were applied, there was no response measured at the collector electrode. This behavior was not unexpected as application of negative voltage attracts the positive response ions directly to the 13 mm diameter, 30 mm tall collar surrounding the flame jet, whereas application of positive voltage greater than a few volts repelled the positive response ions downward to the grounded flame jet.

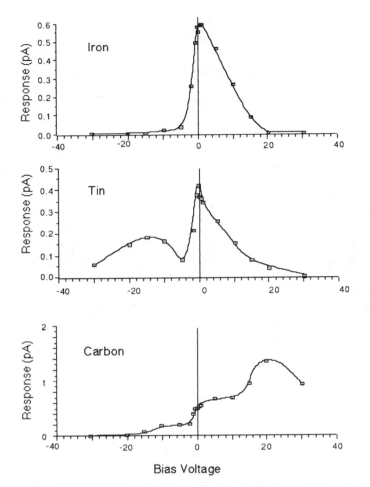

Figure 5.4. HAFID response as a function of voltage applied to the internal collar. Test compounds: 2 pg as Fe from ferrocene; 570 pg as Sn from tetrabutyltin; 124 ng as C from tetradecane. GC column: 30 m × 0.25 mm DB-1. Carrier gas: hydrogen at 0.9 mL/min through the column with a split ratio of 1:10. Oven temperature program: 155 °C initial; hold for 1.5 min; ramping 30 °C/min to 225 °C; hold for 1 min. Detector conditions: H_2 gas flow 335 mL/min doped with 35 ppm SiH_4; jet tip flow of 50 mL/min of O_2 and 115 mL/min of N_2; sheath gas flow 650 mL/min; and the collector electrode was located 70 mm above the flame with a −90 V potential.

Application of a small negative voltage apparently caused the collar to act as an extraction lens for the response ions for organometallics, giving a slight rise in current collected. Response observed for hydrocarbons with application of moderate negative voltages to the collar resulted in a drop in response, possibly due to the direct neutralization of the positive response ions on the surface of the collar. The increase in the response for hydrocarbons upon application of moderate positive voltage may be due to repulsion of the positive response ions upward, aided by the mass flow of the flame effluents toward the negatively biased collector electrode producing a greater response. As a general result, no analytically useful responses were observed as a result of applying an electrical bias to the inner collar of the HAFID.

5.3.2. Wall-Biasing Experiments

The effects of bias electrode placement and various applied voltages upon the HAFID response for a pure hydrocarbon are shown in Figure 5.5. This plot graphically illustrates the importance of electric field interactions upon the mechanism of selective response. In general, the application of high negative potential drew all of the positive background and response ions directly to the bias ring, whereas the application of large positive potentials caused the positive ions to seek ground at either the stainless steel walls of the detector or, if the bias ring was in a low position, at the flame jet tip.

When the bias electrode was positioned within 15 cm of the flame, the response to hydrocarbon compounds in the HAFID increased by application of moderate (up to 50 V) positive or negative potentials applied to the electrode. Negative potentials appeared to act as an extraction lens, drawing the positive response ions upward to be influenced by the field gradient from the −90 V negatively polarized collector electrode. Application of positive potential increased the response to hydrocarbons by deflecting the positive response ions from a collision course with the walls, driving them into the center of the detector, where more of them survived to reach the collector electrode.

When the bias electrode was positioned between 48 and 64 mm above the jet tip, somewhat below the 70 mm height of the collector, the application of bias had very little effect upon the response for hydrocarbons, except in the recurring cases of highest voltages, which repulsed or attracted virtually all of the ions. This behavior may be due to the happenstance of the bias electrode position. Moderate negative potentials were overshadowed by the collector electrode field, whereas application of positive potentials showed a very slight enhancement of response, possibly due to the focusing of the positive response ion in the center of the drift region.

The maximum effect upon response for hydrocarbons was observed when the bias electrode was placed between 64 and 79 mm above the jet tip such that it surrounded the collector electrode. At this position, coulombic interactions between the fields generated by the bias electrode and the collector electrode completely dominated the response of the HAFID. Negative potentials applied to the bias electrode were seen to extract ions from below,

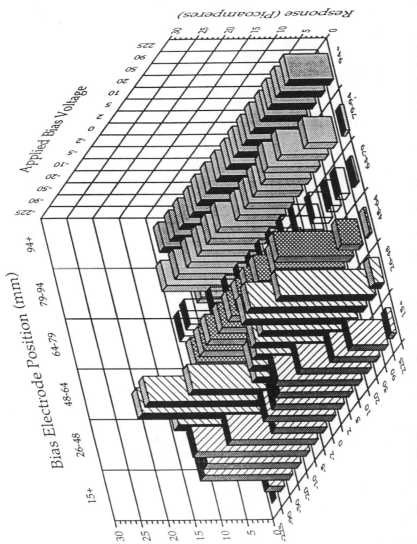
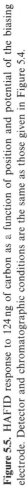

Figure 5.5. HAFID response to 124 ng of carbon as a function of position and potential of the biasing electrode. Detector and chromatographic conditions are the same as those given in Figure 5.4.

forcing a competition with the collector electrode for the ions and causing the reduced (relative to no bias) response. Positive potentials apparently deflected the approaching positive ions to such a degree that the negative field around the collector electrode was shielded from the approaching ions, resulting in a reduced observed response.

Elevating the bias electrode above the collector electrode to greater than 78 mm had a somewhat similar, but lesser, effect upon the response. This was attributed to decreased perturbation of the collector electrode field by the field generated at the bias electrode. Application of negative bias at the 78–94 mm position showed a light extraction effect, whereas positive voltages of 50 and 90 V acted as a focusing element. The highest position of the bias electrode showed a gradual increase in ion current for the extraction effect of the negative bias voltages.

Figures 5.6 and 5.7 depict the response of the HAFID to the iron and tin organometallic components of the test mixture, respectively. Since the general response trends for both metals were similar, this discussion applies to both figures. At the three lowest bias electrode positions, application of slight positive bias acted as an ion focus lens, accelerating the ions inward and resulting in a slight gain in the response measured at the collecting electrode. Application of small negative bias voltages resulted in some response ions becoming neutralized at the bias electrode, although the response collected at the electrode 70 mm above the flame did not diminish significantly.

As with the response to hydrocarbons, application of high potentials of either polarity caused the response ions to be deflected or neutralized, thereby causing large reductions in the ion current measured. With bias electrode at the 48–64 mm position, a lessened sensitivity to changes in applied bias potential was observed. This is similar to the response observed at this electrode position for hydrocarbon compounds.

When the bias electrode was raised to the 64–79 mm position, an increase in the ion current collected was observed. The field at this bias electrode position appeared to impose a focusing effect upon the direction of the response ions, allowing a greater number to remain in the center region of the detector, where the negatively polarized collector electrode imposed its attracting negative field.

With the collector electrode at the 79–94 mm position, a further gain in response was observed, presumably for the same reasons as already described. At positive voltage was applied to the bias electrode, a focusing effect was observed. The maximum response for tin was observed at a bias voltage of +90 V, whereas that for iron was at +50 V.

Response for iron-containing compounds when the bias electrode was raised to the cap of the detector was seen to vary somewhat, owing to the slight focusing and competition effects from the extremes of applied bias voltages.

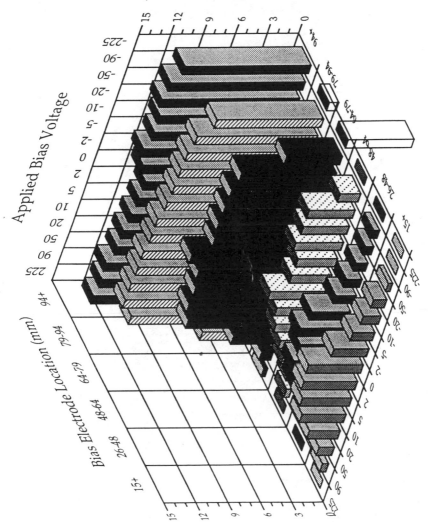

Figure 5.6. HAFID response to 2 pg of iron as a function of biasing electrode position and potential. Detector and chromatographic conditions are the same as those given in Figure 5.4.

137

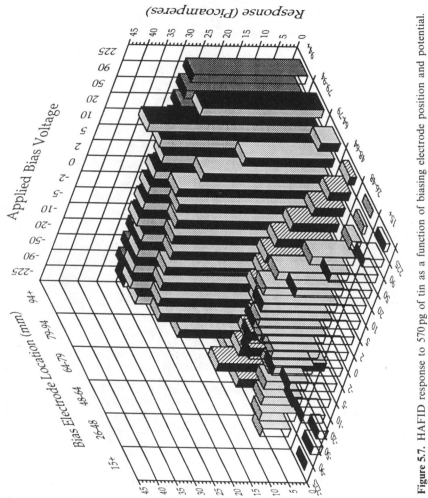

Figure 5.7. HAFID response to 570 pg of tin as a function of biasing electrode position and potential. Detector and chromatographic conditions are the same as those given in Figure 5.4.

138

While not visible in Figure 5.7, the response for tin was not affected at this electrode position, except for the $-225\,V$ application which diminished response by competing with the $-90\,V$ collector electrode.

5.3.3. Computer Modeling Experiments

In an effort to understand the effect of the bias voltage on the response ions, computer modeling of the electric fields inside of the HAFID was performed on a Macintosh personal computer with the MacSimion™ ion simulation program. The output from this program, shown in Figures 5.8 and 5.9, depicts the calculated internal electric field equipotential lines derived from scale drawings of the detector with and without an applied bias potential.

Moving from Figure 5.8a to 5.8e, the location of the biasing electrode was moved up the detector wall corresponding to the locations similar to those employed in Figures 5.5–5.7. Figure 8f depicts the detector in its unbiased configuration. A bias voltage of $+90\,V$ was used for the modeling calculations.

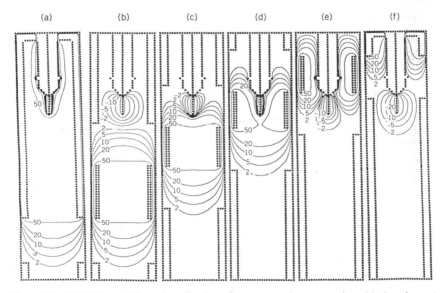

| (a) | (b) | (c) | (d) | (e) | (f) |

Figure 5.8. SIMION modeling of application of constant voltage at various biasing electrode heights. A $+90\,V$ applied to the bias electrode locations similar to those used in the response surfaces shown in Figures 5.5, 5.6 and 5.7. The collector electrode was fixed at 70 mm above the flame jet with a potential of $-90\,V$. Positions of the bias electrode are as follows: (a) 26 + mm above the flame (i.e., the whole detector chimney is biased); (b) 26–48 mm; (c) 48–64 mm; (d) 64–79 mm; (e) 79–94 mm; and (f) 94 + mm (voltage applied to the detector cap). Equipotential lines are drawn at $+50$, $+20$, $+10$, $+5$, $+2$, -2, -5, -10, -20 and $-50\,V$.

The lines drawn in the figure are equipotential surfaces. Note that when the biasing electrode is in the vicinity of the collector electrode, the equipotential surfaces become compressed, making individual potential surfaces difficult to identify. The equipotential lines shown in Figure 8d correspond to the location of the biasing electrode where the greatest effect upon the response of organometallic and hydrocarbon compounds was observed.

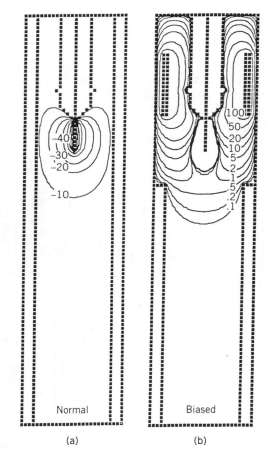

Figure 5.9. Comparison of normal and biased operation of the HAFID by SIMION modeling of electrical fields. (a) The normal HAFID operation with no bias voltage and a − 90 V collecting electrode located 70 mm above the flame jet. Equipotential lines are drawn at − 10, − 20, − 30, − 40, − 50, − 60, − 70 and − 80 V. (b) The biased HAFID. Position and potential of the collecting electrode are the same as in part (a). Position of the + 120 V biasing electrode is 79–94 mm above the flame. Equipotential lines are drawn at + 0.1, + 0.2, + 0.5, + 1, + 2, + 5, + 10, + 20, + 50, and + 100 V.

Figure 5.9 compares the equipotential lines calculated for both the non-biased (normal) HAFID and the biased HAFID. In Figure 5.9a, the normal HAFID, the lines of constant potential are all negative since the only source of voltage is at the $-90\,V$ collector electrode. The lines drawn in Figure 5.9b, the biased HAFID, are all positive potential due to the $+120\,V$ bias. In the vicinity of the $-90\,V$ collector electrode, there exists a small region of compressed negative electrical gradient like the region observed in Figure 5.9a, but the program was incapable of depicting more than 10 equipotential lines in one figure.

It is important to note the differences in the polarity of the fields encountered by the ions. In the normal HAFID shown in Figure 5.9a, the ions drifting toward the collector would experience virtually no field for approximately the first 40 mm of drift; then they would encounter a steadily increasing negative field, attracting the ions toward the $-90\,V$ polarized collecting electrode. The ions in the biased configuration, shown in Figure 5.9b, also drift in a field-free region for approximately 40 mm, but the field they first encounter would be of positive polarity, which opposes and deflects the drift of these positive ions away from the collector electrode and toward ground at the walls of the detector. Any ion that had sufficient momentum to overcome this barrier would pass into the small field-free region leading into the compressed negative potential well of the collecting electrode.

5.4. CONCLUSION

When a positive bias voltage is placed on the wall of the detector in the vicinity of the collecting electrode, an order of magnitude increase in the selectivity of the HAFID toward organometallic compounds can be achieved. This increase in selectivity occurs with some loss of sensitivity. Nevertheless, at the point of near-total suppression of response to hydrocarbons, the response to organometallic compounds is maintained at a high level of sensitivity. Thus, this method of operation has significant analytical utility with respect to the detection of trace quantities of organometallic compounds in the presence of a hydrocarbon background.

Moreover, the experiments described in this chapter are consistent with the collapsing cone mechanistic model of the HAFID. As shown in Figure 5.10, the effect of the biasing voltage was an electric field that repelled a portion of all the response ions, thus producing a loss in sensitivity to all compounds. The greatest amount of ion deflection occurs for the lighter, more mobile hydronium ions (the response ions for hydrocarbons), thereby enhancing the relative collection efficiency of the heavier, less mobile silica cluster ions (the

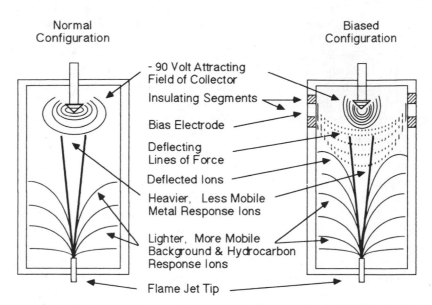

Figure 5.10. Collapsing cone mechanism for organometallic response in the HAFID with biasing electrode.

response ions for organometallic compounds) to produce an overall increase in response selectivity for organometallic compounds.

REFERENCES

1. W. A. Aue and H. H. Hill, Jr., *J. Chromatogr.* **74**, 319 (1972).

2. H. H. Hill, Jr. and W. A. Aue, *J. Chromatogr.* **122**, 515 (1976).

3. M. D. DuPuis and H. H. Hill, Jr., *Anal. Chem.* **51**, 292 (1979).

4. D. R. Hansen, C. H. Lillie, and H. H. Hill, Jr., *J. Chromatogr. Sci.* **23**, 208 (1985).

5. J. E. Roberts and H. H. Hill, Jr., *J. Chromatogr.* **176**, 1 (1979).

6. D. G. McMinn, R. L. Eatherton, and H. H. Hill, Jr., *Anal. Chem.* **56**, 1293.

7. M. M. Gallagher, D. G. McMinn, and H. H. Hill, Jr., *J. Chromatogr.* **518**, 297 (1990).

8. C. H. Lillie, D. G. McMinn, and H. H. Hill, Jr., *Int. J. Mass Spectrom. Ion Processes* **103**, 219 (1991).

9. M. M. Gallagher and H. H. Hill, Jr., *J. High Resolut. Chromatogr.* **13**, 694 (1990).

CHAPTER

6

ELEMENT-SELECTIVE GAS CHROMATOGRAPHIC DETECTION BY ATOMIC PLASMA EMISSION SPECTROSCOPY

PETER C. UDEN

*Department of Chemistry,
University of Massachusetts,
Amherst, Massachusetts 01003-4510*

6.1. INTRODUCTION

6.1.1. The Selective Chromatographic Detector

Since the inception of gas chromatography (GC), the development of detection devices for qualitative and quantitative determination of eluted components has paralleled the advances in separatory efficiency of the chromatographic column. One group of detectors, *bulk property detectors*, respond to changes produced by eluates in a property of the mobile phase itself. Such detectors as the thermal conductivity detector, while approaching universality in detection, suffer from limited sensitivity and inability to characterize the eluate species. In contrast, *solute property detectors* directly measure some physicochemical property of the eluate species. Depending upon the property sensed, they can show high sensitivity and also provide eluate identification and characterization.

Some solute property detectors give "selective" information or, under optimal circumstances, "specific" information on the eluates. Spectroscopic detectors, by virtue of the different spectral properties measured, may be *element selective, structure or functionality selective*, or *property selective*.[1]

From the spectroscopic point of view, the chromatographic sample is distinctive in that analytes are changing in nature and with time; in high-resolution chromatographies, these changes occur very rapidly and thus rigorous demands are placed on the analytical detection system. Effective analysis must involve optimization of both the separation and the detection

Selective Detectors, edited by Robert E. Sievers. Chemical Analysis Series, Vol. 131.
ISBN 0-471-01343-9 © 1995 John Wiley & Sons, Inc.

processes, as well as the interface between them. The most widely used spectroscopic detector in gas chromatography is the mass spectrometer (MS), but infrared (IR) and atomic emission spectroscopy have established a valuable presence as well.

6.1.2. The Element-Selective Detector

Frequently used element selective GC detectors include the alkali flame ionization nitrogen–phosphorus detector (NPD), the flame photometric detector (FPD), selective for sulfur and phosphorus or other elements, and the Hall electrolytic conductivity detector (ELCD), for halogen, nitrogen, and sulfur. The widespread adoption of these detectors shows the value of element-selective detection, but they are each limited in elemental scope in comparison with a full spectral technique such as atomic emission. Multielement chromatographic detection is an important complement to molecular and structural specific detection by interfaced mass spectroscopy and Fourier transform infrared spectroscopy. Atomic emission spectroscopy (AES) is a natural choice for such interfaced detection in view of its capacity in monitor all elements, and the wide employment of plasma excitation sources in AES has focused the efforts of chromatographers to employ its capacities in on line detection.

The major objectives of interfaced chromatography–atomic plasma emission spectroscopy (C/APES) are (a) to monitor eluates for their elemental composition with high elemental sensitivity, (b) to determine the selected element with high selectivity over coeluting elements, (c) to tolerate imcomplete chromatographic resolution from complex matrixes, (d) to detect molecular functionality through derivatization with element-tagged reagents, and (e) to detect a number of elements simultaneously for empirical and molecular formula determination.

6.2. OVERVIEW OF INSTRUMENTATION

6.2.1. Types of Atomic Plasma Emission Chromatographic Detector

The most important plasma emission sources that have been used for gas chromatographic detection have been the microwave-induced and -sustained helium plasma (MIP), and the direct current argon plasma (DCP). Alternating current plasma (ACP), capacitively coupled plasma (CCP), and radio-frequency-induced plasma (RFP) have also been successful. The inductively coupled argon plasma (ICP) has been little used for GC but is finding favor as a mass spectral ion source in the GC–ICPMS mode.

6.2.1.1. The Microwave-Induced Electrical Discharge Plasma (MIP) Detector

In the MIP, an argon or helium plasma is initiated and sustained in a *microwave cavity* that focuses power from a microwave source, typically at 2.45 GHz, into a sample contained in a discharge cell made of quartz or other refractory material.[2,3] Power levels for analytical microwave plasmas are usually around 50–100 W, but power densities are similar to those seen in the ICP or DCP owing to the smaller size of the MIP. High electron temperaturres are produced, notably in the helium plasma, giving intense spectral emission for many elements (including nonmetals) that have poor sensitivity in the argon ICP or DCP. Risby and Talmi compared microwave cavities in their general review of GC/MIP.[4]

The 1/4 wavelength cavity described by Fehsenfeld et al.[5] has been widely used for reduced pressure helium or argon plasmas. In contrast, cavities based upon the TM_{010} cylindrical resonance cavity developed by Beenakker[6] have been most widely adopted for atmospheric pressure GC/MIP. In the Beenakker cavity, light emitted is viewed axially, not transversely through cavity walls whose properties change with time. Atmospheric pressure operation greatly simplifies GC detection, and the configuration allows capillary columns to penetrate within a few millimeters from the plasma, thereby reducing peak shape degardation in the interface.

The performance of the GC/MIP has been improved with a threaded tangential flow torch (TFT),[7,8] to give a self-centering plasma which provides enhanced emission and better stability. The plasma loses relatively little energy to the walls, thus atom formation and excitation are enhanced by comparison with the straight capillary torch. A disadvantage however is the high volume (liters per minute) of helium flow gas required.

The general term GC/AED (atomic emission detection) has become widely used with respect to the microwave-plasma-based system described by Quimby and Sullivan.[9,10] This commercially available instrument, which is being used increasingly in recent years, incorporates a plasma that is generated in an atmospheric pressure flow of helium, composed of GC capillary column flow and a makeup of 40–200 mL/min. It is sustained in a thin-walled silica discharge tube within a water-cooled *microwave reentrant cavity*, which is a circularly symmetric cavity having its central portion narrower than the periphery. Power at around 50 W is supplied by a magnetron and coupled through a waveguide, mitigating the need for microwave tuning. Additional low concentrations of various reagent gases such as hydrogen and oxygen may be added to optimize performance for different elements. A purge flow in the exit chamber of the cavity eliminates deposits on the viewing window of the spectrometer, which uses a movable photodiode array in a flat focal

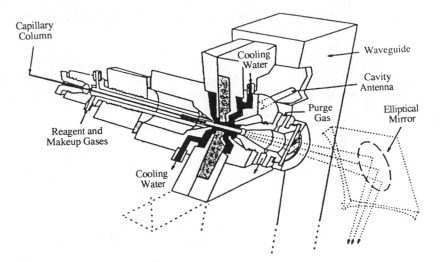

Figure 6.1. Cutaway view of cooled microwave cavity. Reproduced by permission from Sullivan and Quimby.[9] Copyright 1989, Dr. Alfred Huethig Publishers.

plane spectrometer. Figure 6.1 shows a cutaway view of this microwave cavity.

6.2.1.2. Other Low-Power Plasma Chromatographic Detectors

The Surfatron, another microwave-powered atmospheric pressure plasma cavity, operates by surface microwave propagation along a plasma column that may be viewed axially or trransversely, since it extends outside of the plasma structure.[11] Parameters including line selection, pressures of helium and added reagent gases, and tube configuration have been optimized for nonmetallic element detection to give detection limits in the 1–40 pg/s range.[12] Good agreement with electron capture detection (ECD) data was obtained for determinations of pesticides in river water.[13] Factorial analysis and response surface techniques have also been used in Surfatron performance optimization for halogens, and forward power and plasma gas flow rates were found to be most important.[14]

A Fourier transform spectrometer was interfaced to a Surfatron MIP to obtain multielement determination through its multiwavelength capability.[15] The near-IR region covering the range 680–1080 nm was selected for simultaneous detection of fluorine oxygen, chlorine, and carbon because of high halogen sensitivity, simplicity of background, and absence of interferences. Detection limits, however were poorer than with single-element detection with a dispersive system.

A 60 Hz alternating current helium plasma (ACP) acts as a stable, self-maintaining emission source that needs no external initiation and does not extinguish under high solvent loads.[16] The ACP operates as a microarc at power line frequency and may be the simplest and least expensive atomic source to construct and operate. A uniform discharge is generated across two electrodes with an output of ca. 10,000 V at 23 mA.

Atmospheric pressure capacitively coupled plasma (CCP) emission discharges have been developed,[17,18] and their small volume has enabled their use for capillary GC. Such systems can sustain a stable plasma over a wide range of input power (10–500 W) and frequencies (200 kHz to 30 MHz) and carrier gas flow rates down to 20 mL/min.[19]

Rice et al. have reported a series of studies on low-frequency, high-voltage electrodeless discharges sustained in argon, nitrogen, and helium, showing analytically useful emission from the atmospheric pressure afterglow region for GC detection.[20] Since chromatographic effluents are introduced directly into the afterglow discharge region, extinguishment and contamination of the primary discharge are eliminated. With a 45 W discharge at a frequency of 26–27 kHz, F, Cl, Br, I, C, P, S, Si, Hg, and As were determined with limits of detection from 0.5 to 50 pg, linear ranges of $1-7 \times 10^3$ and selectivities from 20 (As) to 7000 (Hg).

Skelton et al. developed a radio-frequency plasma detector (RFPD) operating at power levels of 50–80 W at ca. 335 kHz for sulfur-selective capillary GC of fossil fuels.[21] Sulfur selectivity at 921 nm greater than 10^3 was obtained in petroleum distillates and coal extracts, and a detection limit of 0.5 pg/s with 4 decades linear range was reported.

A stabilized capacitive plasma (SCP) operating with a radio frequency (RF) at 27.12 MHz and with up to 200 W of RF energy has given high long-term stability and minimal matrix interferences.[22] The system incorporates a SCP generator and torch and a fiber-optic link to a filter spectrometer, all simply mounted on the gas chromatograph. The near-IR spectral range from 800 to 1050 nm has given the best performance with detection limits in the low pg/s range for C, Cl, S, and Br.[23] Figure 6.2 shows simultaneous element-specific chromatograms for a range of pesticides containing various heteroatoms.

6.2.1.3. Higher Power Argon Plasma Detectors

The direct current argon plasma (DCP) discharge is sustained by a continuous DC arc and stabilized by flowing inert gas.[24] For chromatographic detection, a cathode jet is placed above two symmetrically placed anode jets in an inverted Y configuration.[25] A 'thermal pinch' gives an arc column having a power level between 500 and 700 W at an operating potential of 40–50 V.

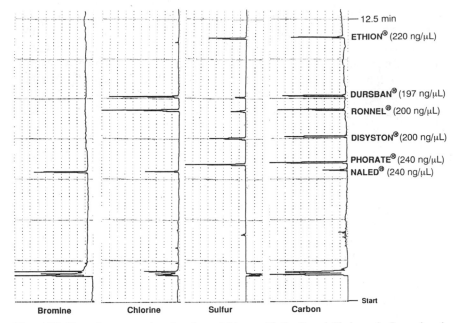

Figure 6.2. Detection comparison for six pesticides on Cl, Br, S, and Cl channels. Reproduced by permission from Knapp et al.[22] Copyright 1990, Gordon & Breach Science Publishers.

The DCP has been interfaced with both high-performance liquid chromatography (HPLC) and GC. Since, unlike the MIP, the DC plasma is not constrained in a discharge tube, additional makeup argon must be introduced annularly around the chromatographic column effluent stream to direct it into the excitation region without loss of chromatographic efficiency.

The inductively coupled argon plasma (ICP) discharge is the most widely used general analytical emission spectrochemical source;[26] it arises from interaction of a radio-frequency field—usually 27 or 41 MHz on argon, flowing through a quartz tube within a copper coil. The RF field creates a varying magnetic field in the gas that generates a stable, spectrally intense plasma discharge with temperatures as great as 9000 K. The ICP is a natural complement for liquid chromatography, and HPLC/ICP procedures have been quite widely adopted. It has been less used as a GC detector but may be useful for those elements, particularly metals, the sensitivity of which is high in the argon plasma discharge.

6.2.2. Analytical Information from Chromatography–Atomic Emission Detection (C/AED)

The detection limit and sensitivity for an element in the AED depends on its spectral intensity at the measured wavelength and the signal-to-noise ratio

(S/N). Each element has many emission wavelengths, and the best must be chosen for analytical purposes. Sensitivity, defined by the slope of the response curve, is less often used in C/AED than are detection limits, given as an absolute value of element mass (in a resolved peak) or in mass flow rate units.

Interelement selectivity depends on emission properties of the element and of possible interferences, and on the resolution of the spectroscopic

Table 6.1. Figures of Merit for Atmospheric Pressure Helium MIP GC Detection

Element	Wavelength (nm)	Detection Limit[a] [pg/s (pg)]	Selectivity vs. C	LDR[b]
Carbon[c]	247.9	2.7(12)	1	>1,000
Carbon[d]	193.1	2.6	1	21,000
Hydrogen[c]	656.3	7.5(22)	160	500
Hydrogen[d]	486.1	2.2	Variable	6,000
Deuterium[c]	656.1	7.4(20)	194	500
Boron[c]	249.8	3.6(27)	9,300	500
Chlorine[d]	479.5	39	25,000	20,000
Bromine[d]	470.5	10	10,400	>1,000
Fluorine[d]	685.6	40	30,000	2,000
Sulfur[d]	180.7	1.7	150,000	20,000
Phosphorus[d]	177.5	1	5,000	1,000
Silicon[d]	251.6	7.0	90,000	40,000
Oxygen[d]	777.2	75	25,000	4,000
Nitrogen	174.2	7.0	6,000	43,000
Aluminum[d]	396.2	5.0	>10,000	>1,000
Antimony	217.6	5.0	19,000	>1,000
Gallium[d]	294.3	ca. 200	>10,000	>500
Germanium[c]	265.1	1.3(3.9)	7,600	>1,000
Tin[c]	284.0	1.6(6.1)	36,000	>1,000
Tin[d]	303.1	(0.5)	30,000	>1,000
Arsenic[d]	189.0	3.0	47,000	500
Selenium[d]	196.1	4.0	50,000	>1,000
Chromium[d]	267.7	7.5	108,000	>1,000
Iron[d]	302.1	0.05	3,500,000	>1,000
Lead[c]	283.3	0.17(0.71)	25,000	>1,000
Mercury[d]	253.7	0.1	3,000,000	>1,000
Vanadium[d]	292.4	4.0	36,000	>1,000
Titanium[d]	338.4	1.0	50,000	>1,000
Nickel[d]	301.2	1.0	200,500	>1,000
Palladium	340.4	5.0	>10,000	>1,000
Manganese[d]	257.6	1.6(7.7)	110,000	>1,000

[a]Detection limit: three times the signal-to-noise ratio (S/N).
[b]LDR: linear dynamic range.
[c]Conventional TM_{010} MIP, University of Massachusetts.
[d]Hewlett Packard 5921A, Hewlett Packard or University of Massachusetts.

measurement system. A useful definition is the peak area response per mole of analyte element divided by the peak area response of the "background" element, per mole of that element.

Dynamic measurement range of response in capillary GC/AED typically extends from the upper linear analyte capacity, ca. 100 ng, to the detection limit of the target element, which is typically in the pg range.

Carbon-selective detection may be thought of as a universal mode of detection for organic compounds. The emission wavelength most often used is from a carbon ion line at 247.9 nm, but other spectral emission features corresponding to CN, CH, etc. may be used.[4] This AED mode response is analogous to flame ionization but is more completely independent of carbon atom environment and exhibits as great or greater sensitivity.[3]

Table 6.1 summarizes elemental detection limits, selectivities, and linear dynamic ranges for atmospheric pressure microwave-induced helium plasma capillary GC detectors.

6.2.3. Plasma Interfacing with Capillary GC

Since eluent from GC columns is at atmospheric pressure, interfacing is simpler for atmospheric pressure plasmas than for reduced pressure plasmas. For the latter, interfacing involves evacuating a sample chamber within the MIP cavity to a pressure of ca. 1 torr.[27] For packed columns, little degradation occurs in peak efficiency, but the volume of the chamber leads to some degradation of capillary peak efficiency. The atmospheric pressure cavities such as the TM_{010} are very easy to interface with capillary GC columns since these can be brought to within a few millimeters of the plasma, giving minimal transfer volume.[28] Helium makeup gas or other reactant gases can be introduced within the transfer line to optimize plasma performance and minimize peak broadening. The microwave induced plasmas (MIP) have found much greater use in GC than in HPLC interfacing, although the application of the direct injection nebulizer (DIN) for microbore column effluent flow rates may increase the method's potential.[29]

6.3. GAS CHROMATOGRAPHY–ATOMIC EMISSION DETECTION

6.3.1. GC/AED of Nonmetallic Elements

It is clear that although many of the more impressive applications of GC/AED may be in the areas of inorganic and organometallic application, as noted subsequently, it is essential that a system provide good analytical performance for nonmetals and particularly organic heteroatom elements such as the halogens, sulfur, nitrogen, and oxygen.

The microwave-induced helium plasmas (MIP) have been most widely used for detection of nonmetals since for most of these elements argon metastable energy carriers show insufficient collisional energy transfer for adequate excitation.

6.3.1.1. Reduced Pressure Plasmas

McCormack et al.[30] and Bache and Lisk[2] devised reduced pressure argon GC/MIP systems for selective detection of P, S, F, Cl, Br, I, and C, obtaining detection limits between 10^{-7} and 10^{-12} g/s, but with poor selectivities against carbon. Bache and Lisk later obtained improved sensitivities for the halogens, P, and S, using a reduced pressure helium MIP.[31] Line emission was observed in the helium plasma, giving better performance than the less energetic argon MIP, where Cl, Br, N, and O show weak diatomic molecular emission. McLean et al.[3] developed a system using scavenger gas to prevent carbon deposits from forming on the inside of the plasma tube. Detection limits were between 0.03 and 0.09 ng/s for C, H, D, F, Cl, Br, I, and S, with selectivities against carbon between 400 to 2300 nm. Detection limits for O and N were around 3 ng/s. Arsenic and antimony were determined in environmental samples by derivatization and reaction to form stable triphenylarsine and triphenylstibine, measured by reduced-pressure GC/MIP at 228.8 and 259.8 nm, respectively, with detection limits of 20 and 50 pg.[32]

The analytical performance of a commercial instrument utilizing a multichannel polychromator spectrometer with Rowland Circle optics was reported by Brenner.[33] With a helium plasma at 0.5–3.0 torr and oxygen or nitrogen scavenger gas, detection limits were from 0.02 ng/s for Br to 4 ng/s for O, and linearity of 3–4 decades was seen. Empirical formulas for polychlorinated biphenyls (PCBs), mercaptans, and diols were reported within 5–10% at analyte levels as low as 100 ppm. This instrument has been used in various studies either with the low-pressure Evenson-type 1/4 wavelength cavity or with atmospheric pressure TM_{010} (Beenakker) cavities.

Hagen et al.[34] used a multichannel instrument utilizing a polychromator configuration with Rowland Circle optics, as well as a low-pressure Evenson cavity. They used elemental derivatization reagents such as chlorofluoroacetic anhydride to tag analyte molecules, thereby permitting detection of acylated amines through their F and Cl content. With a similar reduced pressure system, Zeng et al. focused upon oxygen-specific detection;[35] the use of purified plasma gases with rigorous exclusion of air gave a limit of detection to 0.3 ng/s with 3 decades linear dynamic range. Oxygen-specific analysis is becoming more and more important with the increasing use of oxygenates in fuel oils, and oxygen fingerprints of such materials have potential in environmental analyses.

Sklarew et al.[36] observed that although the low-pressure Evenson source provides the greatest excitation energy for nonmetals, its sensitivity is inherently less than that of the Beenakker cavity because of its transverse viewing geometry. The atmospheric pressure Beenakker cavity provides a more efficient way to couple microwave energy into the plasma, and its axial geometry provides inherently more sensitivity; however, its high pressure causes tube erosion, thus resulting in reliability problems. Sklarew et al. devised a modified Beenakker cavity operating at 20 torr, which was effective for sulfur-specific detection, as well as empirical formula determination in complex matrixes such as an oil shale retort off-gas. Hooker and DeZwann applied the polychromator system with the Evenson cavity in a configuration

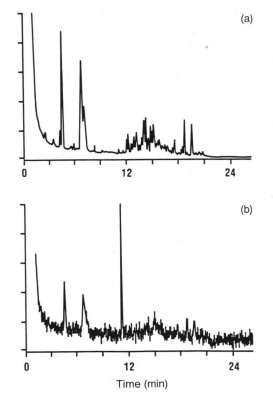

Figure 6.3. (a) Carbon and (b) fluorine MIP emission chromatogram of an extracted sample of blood plasma from a rat dosed with a fluorinated compound. Reproduced by permission from Hooker and DeZwaan.[38] Copyright 1989, American Chemical Society.

that permitted simultaneous atomic emission and mass spectrometric detection. Chromatographic effluent was split within the GC oven at either reduced pressure[37,38] or atmospheric pressure.[39] Figure 6.3 shows carbon and fluorine responses for an extracted rat blood plasma sample. The potential of deuterium-specific detection was also demonstrated in blood plasma determination of a rat dosed with deuterium-enriched benzodiazepin.[37]

Olsen et al.[40] compared reduced pressure and atmospheric pressure MIP systems for Hg, Se, and As detection in organometalloids in shale oil matrices; they found the latter to be superior both in detection limits and in selectivities.

6.3.1.2. Atmospheric Pressure Plasmas

More efficient transfer of microwave power to the plasma by means of such cavities as the Beenakker TM_{010} allows plasmas to be maintained at atmospheric pressure at low power levels similar to those of reduced pressure cavities. Also, light emitted from the plasma can be viewed axially rather than transversely through a quartz discharge tube wall subject to variations in spectral throughput because of contamination and corrosion from extended use.

Flexible fused silica capillary (FSOT) columns can be interfaced to within a few millimeters of the plasma, and such direct interfaces have been widely used in capillary GC/MIP, although three are more advantages in incorporating a gas switching device to introduce additive gases or purge segments of the chromatographic eluates.

A number of developments show the versatility of GC/MIP research and applications; widespread adoption of the technique for more routine use is now feasible with the commercial introduction of an instrumental system mentioned earlier.[9,10] Initial reports have focused upon C, H, N, and O detection including sensitivity, precision, peak shapes induced by the plasma, and response variation between compounds. Precision ranged from 2% (C, H, N) to 7% (O), and empirical formulas could be calculated with an accuracy to one atom in most cases.

For halogen detection, the MIP has advantages over the other major selective GC detectors used, the electron capture detector (ECD) and the Hall electrolytic conductivity detector (ELCD). Although the MIP does not have as great sensitivity as the ECD for polyhalogenated compounds, it has the advantage of almost uniform elemental response for each halogen, irrespective of analyte molecular structure. A major advantage over the ELCD is element specificity for different halogens, which is not shown by the halogen-specific mode of that detector.

Among the capabilities of multichannel AED is its ability to perform quantitative element ratio measurements to obtain empirical formulas of

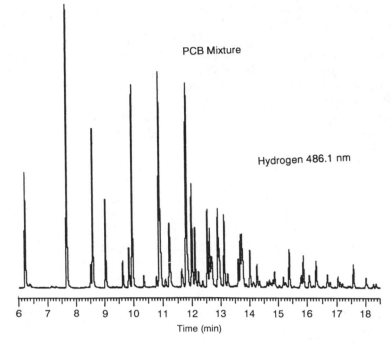

Figure 6.4. GC/AED hydrogen response to a polychlorinated biphenyl (PCB) mixture. Reproduced by permission from Wylie et al.[41] Copyright 1990, Dr. Alfred Huethig Publishers.

eluates.[41] Wylie et al. developed empirical formula calculation algorithms using both internal and external standards. Figure 6.4 depicts a hydrogen-selective chromatogram for a homologous series of PCBs, with each compound containing 12 carbons and a total of 10 hydrogens and chlorines. Good correspondence was seen for ratioed response factors against hydrogen peak height, giving values for chlorine content within 0.2 of an integer up to 5 chlorines in the molecule. Such determinations do not achieve the accuracy of conventional microanalysis, but measurements are made directly on GC eluent peaks for analyte levels up to 6 orders of magnitude smaller!

Kovacic and Ramus[42] evaluated GC/AED elemental response factors for compound independence, C, Cl, F, N, and O showing 3%, 6%, 5%, 6%, and relative standard deviations (RSDs), respectively, for a wide range of aliphatic, aromatic, and heterocyclic compounds.

Oxygen-selective detection with an oxygen-to-carbon selectivity of 10^2 and a thousandfold linearity was found by Bradley and Carnahan[43] with a TM_{010} cavity and a polychromator system. Background oxygen spectral emission from plasma gas impurities, leaks, or back-diffusion into the plasma

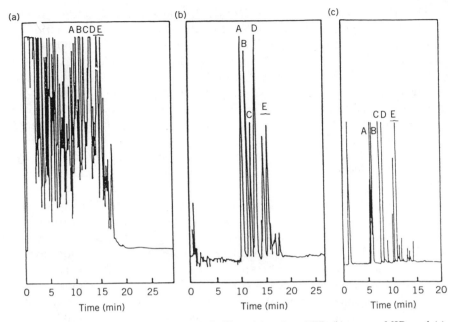

Figure 6.5. Chromatograms of light coal distillate: (a) carbon MIP, (b) oxygen MIP, and (c) FID trace of phenolics concentrate of the same distillate; phenol (A), *o*-chlorophenol (B), *o*-cresol (C), *m*- and *p*-cresols (D), C_2-phenols (E); temperature program, GC/MIP, 30 °C (1 min hold) to 250 °C at 10 °C/min; GC/FID, 50–250 °C at 10 °C/min. Reproduced by permission from Bradley and Carnahan.[43] Copyright 1988, American Chemical Society.

was minimized to give oxygen detection limits from 2 and 500 ppm in complex petroleum distillates. The selective detection of phenols in a light coal liquid distillate is seen in Figure 6.5.

Pedersen-Bjergaard and Greibrokk[44] have evaluated a radio-frequency plasma detector as described by Skelton et al.[45] for oxygen-selective detection. Purification of plasma and purge gases was critical, and interferent oxygen signals deriving from release of the element from the discharge tube by eluting hydrocarbons could be minimized by using methane doping in a 15 W plasma. A 1000:1 oxygen-to-carbon selectivity was obtained with a 100 pg/s detection limit.

Further examples of petroleum analysis using GC/AED are cited by Kosman,[46] who comments on the usefulness of compound-independent calibration wherein a simple compound such as methane can be used to calibrate the detector simultaneously for carbon and hydrogen for application in complex matrixes. The importance of element-selective detection profiles in simulated distillation applications is stressed, and pattern recognition using element rationing to type crudes and oil spills is also emphasized.[47]

GC/AED has been compared with other element-selective detectors, namely, the ECD, ELCD (Hall detector), NPD, and FPD, for the determination of pesticide residues in agricultural products.[48] Twelve agricultural commodities were fortified with 10 commonly used pesticides, which were then extracted without the use of cleanup steps. The AED was used for C, P, Cl, F, N, and S analyses, and its high selectivity allowed analysis in all 12 extracts. By contrast, interferences restricted the ECD use to five commodities, the ELCD to eight, and the FPD and NPD to nine only. For the samples that produced the greatest levels of interferences for other detectors, onions, strawberries, and alfalfa, AED provided the greatest degree of both selectivity and sensitivity, particularly for phosphorus detection. The potential of GC/AED for determinations of such pesticide residues appears to be very great.

Miles and Quimby[49] utilized supercritical fluid extraction (SFE) for sample preparation prior to GC/AED of organosulfur and organophosphorus compounds in various spices. They demonstrated the higher interelement selectivity between S and P than that obtainable with the FPD. The incorporation of four phosphorus emission lines and three sulfur lines into the respective analytical protocols enabled high confidence in the assignments made.

Sandra has shown the utility of GC/AED for a range of environmental air pollution applications.[50] The highly toxic compound carbonyl sulfide was determined in the presence of high levels of hydrocarbons: $100\,\mu L$ vapor samples were injected onto porous-polymer-coated capillary columns for sulfur determination at the ppb level. Sulfur species in city air were determined after concentrating on a thick silicone film trap, thermal desorption, and cryofocusing.

GC/AED has been effectively interfaced with analytical pyrolysis to provide characterization information in organic geochemistry. Rowland et al.[51] applied sulfur-selective pyrolysis/GC/AED to diatomaceous Pleistocene sediments and utilized the data together with MS information to identify both high- and lower-volatility pyrolyzates. Figure 6.6 shows representative GC/AED and GC/MS chromatograms. Samples of the order of 200 mg were pyrolyzed for 20 s at 610 °C in a quartz probe within a platinum coil. This study emphasized the value of parallel GC/AED and GC/MS in such investigations.

The applicability of GC/AED in the complex pyrolyzates produced from organic geochemical samples was further demonstrated by Seeley et al., who carried out a study of Monterey kerogen.[52] Element-specific pyrograms were obtained, and Figure 6.7 shows a typical sulfur-specific pyrogram from a Monterey kerogen pyrolyzed at 800 °C for 10 s. Observed sulfur species have been identified primarily as arylthiophenic species. Figure 6.8 shows the three-dimensional arsenic spectral profile employed in this determination, and Figure 6.9 shows a portion of an arsenic-selective chromatogram, the

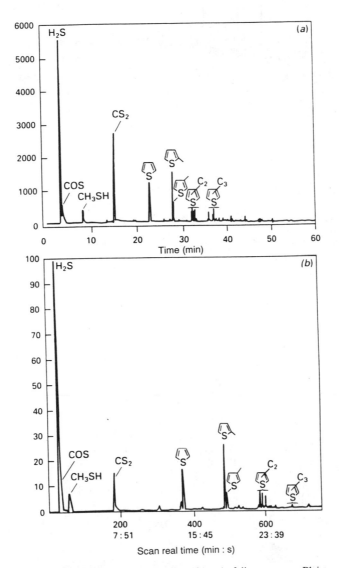

Figure 6.6. (a) Pyrolysis/GC/AED (S emission line 181 nm) of diatomaceous Pleistocene sediment from Peru upwelling zone. (b) Pyrolysis/GC/MS (summed mass fragmentogram m/z 34, 48, 76, 84, 98, 112). Reproduced by permission from Rowland et al.[51] Copyright 1993, The Royal Society of Chemistry.

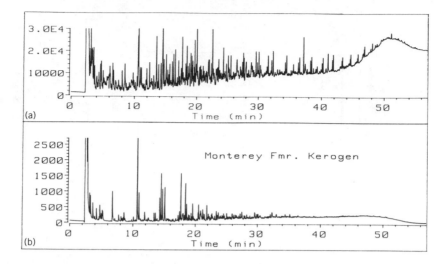

Figure 6.7. (a) Carbon- and (b) sulfur-selective chromatograms (193 and 181 nm, respectively) from Monterey kerogen pyrolyzed at 800 °C for 10 s.

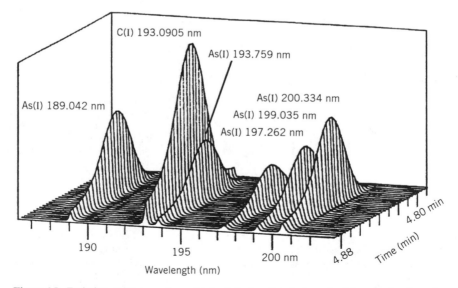

Figure 6.8. Emission spectra vs. time obtained during the elution of triphenylarsine into the Hewlett Packard 5921A plasma.

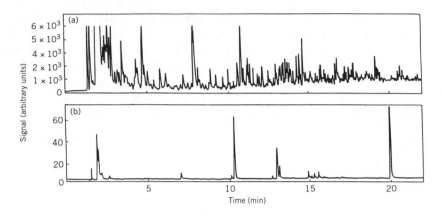

Figure 6.9. (a) Carbon- and (b) arsenic-selective chromatograms (193 and 189 nm, respectively) from Monterey kerogen pyrolyzed at 800 °C for 10 s. Reproduced by permission from Seeley et al.[52] Copyright 1992, The Royal Society of Chemistry.

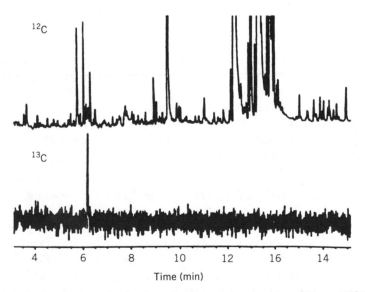

Figure 6.10. Carbon-13 determination by molecular emission detection of 30 pg of [^{13}C]nitrobenzene, spiked into urine and extracted. Reproduced by permission from Sullivan.[56] Copyright 1991, Elsevier Science Publishers.

selectivity afforded by the element-specific protocol used in the analysis, leaving no doubt as to the authenticity of the organoarsenic compounds. An on-line butylation procedure applied to this analytical pyrolysis has also shown the presence of organoselenium species and organophosphorus compounds produced by the kerogen thermal degradation. Tenzer and Heumann[53] have determined organoselenium compounds in aquatic systems, measuring dimethylselenium in seawater to the 50 pg/L level using capillary cold trapping from 50 mL samples.

The ability of GC/AED to differentiate and quantify isotopes of elements of low atomic number such as hydrogen, carbon, nitrogen, and oxygen has considerable potential in the area of stable isotope labeling. Deruaz et al.[54] have used GC/AED with hydrogen, deuterium/sulfur, and iodine detection in a study of the purity of deuteroiodomethane used for derivatization of immunoenhancer drugs. H, D and I can be conveniently determined simultaneously. Quimby et al.[55] and Sullivan[56] reported the ability of GC/AED to determine ^{13}C selectivity over ^{12}C, utilizing the molecular emission from intense C bands in the vacuum ultraviolet region. Figure 6.10 shows ^{13}C and ^{12}C selective chromatograms where [^{13}C]nitrobenzene was spiked into urine at the 3 ng/mL level; 30 pg of the ^{13}C compound is readily detected. The instrumental recipe is set to measure selectivity over ^{12}C taking into account the natural abundance (ca 1%) of ^{13}C. This selectivity is reported as > 2000.

6.3.2. GC/AED of Metals

GC/AED methods are valuable in confirming elution of metallic compounds and in acquiring valuable analytical data, since metal elemental detection is usually more sensitive and selective than elemental selection for nonmetals. Many volatile binary metal compounds, organometallics, and metal chelates can be quantitatively gas chromatographed,[57] although in some instances the absence of confirmatory elution data based on elemental content of eluates has cast doubt on GC authenticity.

6.3.2.1. GC/AED of Main Group Metallic Compounds

As shown in Table 6.1, GC/MIP data have been obtained for many main and transition group metals. Among the former, tin, lead, and mercury have been studied most frequently as they are determinable by GC/MIP with TM_{010} cavities with sub-pg/s detection limits. Lead- and carbon-specific detection methods have been compared for trialkyllead chlorides extracted from an industrial plant effluent and reacted with butyl Grignard reagent to form their trialkylbutyllead derivatives.[58] The level of interference from

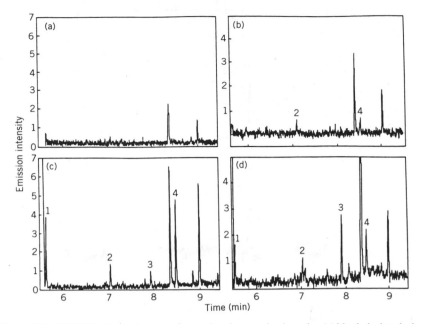

Figure 6.11. GC/AED chromatogram of propylated organolead species: (a) blank during derivatization of pure hexane with propylmagnesium chloride; (b) blank from complete analytical procedure; (c) chromatogram from tap water; (d) chromatogram of a polar snow sample. Peaks: 1, Me_3Pb^+; 2, Me_2Pb^{2+}; 3, Et_3Pb^+; and 4, Et_2Pb^{2+}. Reproduced by permission from Lobinski and Adams.[59] Copyright 1992, The Royal Society of Chemistry.

carbon-containing compounds prevented any qualitative or quantitative determination of the trialkyllead compounds by GC/FID, GC/ECD or GC/MS without extensive cleanup and loss of analyte.

Lobinski and Adams[59] developed speciation of organolead compounds in water at the sub-ng dm^{-3} level. Sample preparation involved extraction of ionic organolead species as dithiocarbamate complexes, followed by their propylation by Grignard reaction. Sample injection to the capillary column was by concentration on a Tenax trap. Figure 6.11 shows a mixture of propylated lead species present in a rainwater sample at the ng/L level.

In a comparison between reduced and atmospheric pressure MIP systems, Olsen et al. found a 1 pg detection limit for mercury for the latter, with selectivity over carbon of 10,000.[40] Emteborg et al.[60] developed a method for simultaneous determination of methylmercury, ethylmercury, and inorganic mercury by dithiocarbamate resin concentration in a flow injection system, followed by butylation. With water sample volumes of 500 mL, a detection limit for methylmercury of 0.05 ng/L was obtained. Figure 6.12 shows response

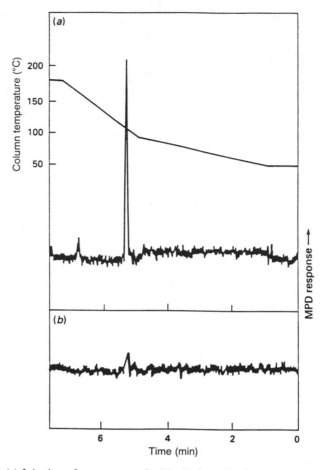

Figure 6.12. (a) Injection of a water sample (20 mL) from the 100 m level, Gulf of Bothnia, Sweden, showing 0.35 ng/L of methylmercury. (b) Blank. Reproduced by permission from Emteborg et al.[60] Copyright 1993, The Royal Society of Chemistry.

response from a 20 µL water sample from the 100 m level in the Gulf of Bothnia, Sweden, showing $0.35 \, \text{ng/L}^{-1}$ of methylmercury.

Lobinski et al. reported parallel derivatization procedures for organotin compounds allowing detection at about 0.05 pg.[61] Tributyltin species have also been determined by GC/AED utilizing soil phase extractive (SPE) concentration from seawater, followed by conversion to the corresponding hydride using solid sodium borohydride reagent, in line to the capillary column.[62] It was found that interferences in the GC/AED chromatogram were produced by volatile organotin compounds deriving from residues in

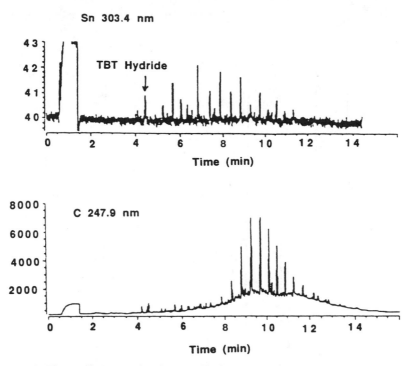

Figure 6.13. Tin-specific (top) and carbon-specific (bottom) chromatograms of Newport, Rhode Island, seawater after solid phase extraction and hydride conversion of tributyltin chloride to tributyltin (TBT) hydride. Reproduced by permission from Dowling and Uden.[62] Copyright 1993, Elsevier Science Publishers.

the SPE cartridge; these could be removed by solvent treatment of the cartridge. Figure 6.13 shows such a chromatogram.

6.3.2.2. GC/AED of Transition Metal Compounds

For GC of volatile metal complexes, most emphasis has been placed upon chelating ligands of 2,4-pentanedione (acetylacetone) and its analogs.[57,63] Black and Sievers[64] used microwave plasma detection to determine chromium as its trifluoroacetylacetonate in blood plasma with excellent quantitation and precision. An example of the utility of GC/MIP detection was in a study of ligand redistribution and reaction kinetics of gallium, indium, and aluminum chelates[65]. Figure 6.14 shows a capillary GC study of redistribution of trifluoroacetylacetone (TFA) (1,1,1-trifluoropentane-2,4-dione) and trifluoro-acetyl isobutanoylmethane (TIB) (1,1,1-trifluoro-6-methylheptane-2,4-dione) ligands on gallium and aluminum atom centers. $Al(TFA)_3$ and $Ga(TIB)_3$

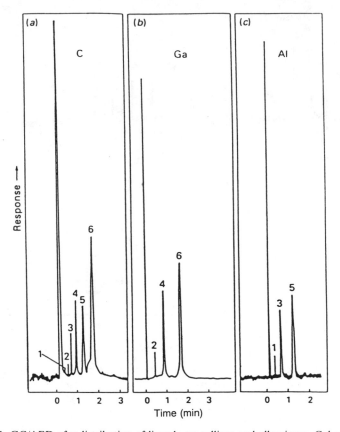

Figure 6.14. GC/AED of redistribution of ligands on gallium and alluminum. Column 5 m × 0.2 mm i.d. FSOT SE 30 capillary at 108 °C: (a) carbon detection at 247.9 nm; (b) gallium detection at 294.3 nm; (c) aluminum detection at 396.1 nm. See the text for identification of the peaks. Reproduced by permission from Uden and Wang.[65] Copyright 1988, Royal Society of Chemistry.

were reacted in chloroform solution for 24 h at 25 °C and the products monitored with carbon-, aluminum-, and gallium-specific GC detection. Individual peaks were identified as follows: 1, $Al(TFA)_2(TIB)$; 2, $Ga(TFA)_2$ (TIB); 3, $Al(TFA)(TIB)_2$; 4, $Ga(TFA)(TIB)_2$; 5, $Al(TIB)_3$; and 6, $Ga(TIB)_3$.

Tetradentate Schiff base ketoiminate complexes of divalent transition metals have proved to have excellent GC characteristics allowing quantitative elution from capillary columns. GC/AED selective chromatograms for copper, nickle, palladium and oxovanadium complexes of N, N'-propylenebis(trifluoro-acetylacetoneimine) are shown in Figure 6.15.[66] The distorted profile of the vanadium-specific chromatogram is due to the presence of stereoisomers of

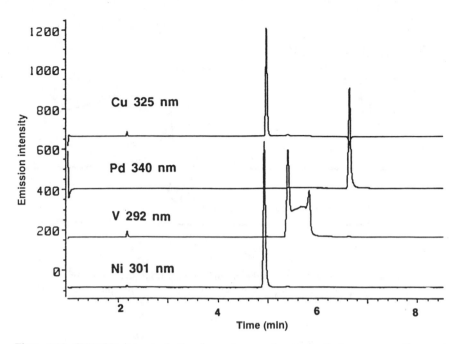

Figure 6.15. GC/AED element-selective chromatogram of copper, palladium, oxovanadium, and nickel derivatives of *N*, *N'*-propylenebis(trifluoroacetylacetoneimine). Reproduced by permission from Uden and Zeng.[66] Copyright 1992, Friedr. Vieweg und Sohn, Pergamon Press.

the oxovanadium chelate for which interconversion occurs on the GC column. Multielement GC/AED for N, C, F, O, and V was used to investigate this phenomenon.

Vanadium-specific detection has also been of value in fingerprinting of metalloporphyrins in crude oils.[67] Figure 6.16 illustrates the selectivity of vanadium detection against a carbon matrix for a fraction extracted from Boscan crude oil. Detection using two independent wavelength recipes at 269 and 292 nm confirms the identity of the vanadium profiles in the two spectral regions.

Many π-bonded organometallics such as metallocenes are well behaved in capillary GC and provide good model compounds for evaluation of GC/MIP detection for their constituent elements. Sensitive and selective detection of iron, cobalt, nickel, chromium, manganese, and vanadium in a series of cyclopentadienyl carbonyl/nitrosyl compounds was obtained, verifying elution of some previously unchromatographed compounds.[68] The manganese organometallics detected have been measured as gasoline additives by GC/AED using a DCP system,[69] and GC/MIP may also be used if solvent removal

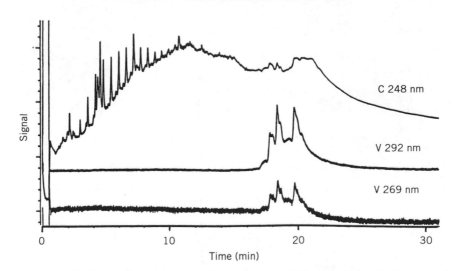

Figure 6.16. Carbon- and vanadium-specific chromatograms from analysis of an extract of Boscan crude oil. Reproduced by permission from Zeng et al.[67] Copyright 1992, Dr. Alfred Huethig Publishers.

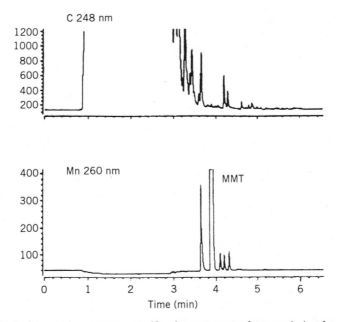

Figure 6.17. Carbon- and manganese-specific chromatograms from analysis of a Canadian gasoline. Reproduced by permission from Zeng et al.[67] Copyright 1992, Dr. Alfred Huethig Publishers.

is carried out following sample injection. An example of such a trace metal GC/MIP analysis is shown in Figure 6.17 for a gasoline sample containing the additive MMT [(methylcyclopentadienyl)manganese tricarbonyl].

The desirable analytical characteristics of metal-specific detection are very clear and in general are limited far more by any shortcomings of GC column behavior than by spectral detection deficiencies.

6.4. THE FUTURE DEVELOPMENT OF ATOMIC EMISSION CHROMATOGRAPHIC DETECTION

In practical terms, the wider adoption of plasma spectral detection will depend on the introduction of more standardized commercial instrumentation to permit interlaboratory comparisons of data and the development of recommended methods of analysis that can be widely used. Many areas of analysis are subject to restrictions designed to ensure high levels of accuracy and precision in analysis. Plasma chromatographic detection has already demonstrated a wide utility in academic, governmental, and industrial laboratories, and the recent commercial introduction of an integrated GC/AED system suggests that the future of this technique is strong. Fully integrated units that circumvent the need for analysts to interface their own chromatograph, emission device, and spectrometer may well become as familiar in the future as GC/MS and GC/FTIR systems are today. Integrated condensed phase chromatographic systems will be longer delayed, but their eventual adoption is highly probable in view of the broad scope of these analytical separation methods.

REFERENCES

1. P. C. Uden, in *Quantitative Analysis Using Chromatographic Techniques* (E. Katz, ed.), p. 99. Wiley, Chichester and New York, 1987.
2. C. A. Bache and D. J. Lisk, *Anal. Chem.* **39**, 786 (1976).
3. W. R. McLean, D. L. Stanton, and G. E. Penketh, *Analyst (London)* **98**, 432 (1973).
4. T. H. Risby and Y. Talmi, *CRC Crit. Rev. Anal. Chem.* **14**(3), 231 (1983).
5. F. C. Fehsenfeld, K. M. Evenson, and H. P. Broida, *Rev. Sci. Instrum.* **36**(3), 294 (1965).
6. C. I. M. Breenakker, *Spectrochim. Acta* **31B**, 483 (1976).
7. A. Bollo-Kamara and E. G. Codding, *Spectrochim. Acta* **36B**, 973 (1981).
8. S. R. Goode, B. Chambers, and N. P. Buddin, *Spectrochim. Acta* **40B**, 329 (1985).
9. J. J. Sullivan and B. D. Quimby, *J. High Resolut. Chromatogr.* **12**(5), 282 (1989).

10. R. L. Firor, *Am. Lab, (Fairfield, Conn.)* **21**(5), 40 (1989).

11. M. H. Abdellah, S. Coulombe, and J. M. Mermet, *Spectrochim. Acta* **37B**, 583 (1982).

12. B. Riviere, J.-M. Mermet, and D. Derauz, *J. Anal. At. Spectrom.* **2**, 705 (1987).

13. S. Coulombe, K. C. Tran, and J. Hubert, *ACS Symp. Ser.* **479**, 189 (1992).

14. M. Caetano, R. E. Golding, and E. A. Key, *J. Anal. At. Spectrom.* **7**, 1007 (1992).

15. C. Lauzon, K. C. Tran, and J. Hubert, *J. Anal. At. Spectrom.* **3**, 901 (1988).

16. R. B. Costanzo and E. F. Barry, *Anal. Chem.* **60**, 826 (1988).

17. B. M. Patel, E. Heithmar, and J. D. Winefordner, *Anal. Chem.* **59**, 2374 (1987).

18. D. C. Liang and M. W. Blades, *Anal. Chem.* **60**, 27 (1988).

19. D. Huang and M. W. Blades, *J. Anal. At. Spec.* **6**, 215 (1991).

20. G. W. Rice, A. P. D'Silva, and V. A. Fassel, *Spectrochim. Acta* **40B**, 1573 (1985).

21. R. J. Skelton, Jr., H.-C. K. Chang, P. B. Farnsworth, K. E. Markides, and M. L. Lee, *Anal. Chem.* **61**, 2292 (1989).

22. G. Knapp, E. Leitner, M. Michaelis, B. Platzer, and A. Schalk, *Int. J. Environ. Anal. Chem.* **38**, 369 (1990).

23. B. Platzer, R. Gross, E. Leitner, A. Schalk, H. Sinabell, H. Zach, and G. Knapp, *ACS Symp. Ser.* **479**, 152 (1992).

24. R. J. Decker, *Spectrochim. Acta* **35B**, 19 (1980).

25. J. O. Beyer, Ph.D. Dissertation, University of Massachusetts, Amherst (1984).

26. R. M. Barnes, *CRC Crit. Rev. Anal. Chem.* **7**, 203 (1978).

27. W. J. Hoskin, *The MPD 850 Organic Analyzer*, Applied Chromatography Systems Ltd., Luton, UK, 1977.

28. S. A. Estes, P. C. Uden, and R. M. Barnes, *Anal. Chem.* **53**, 1829 (1981).

29. K. E. LeFreniere, V. A. Fassel, and D. E. Eckel, *Anal. Chem.* **59**, 879 (1986).

30. A. J. McCormack, S. C. Tong, and W. D. Cooke, *Anal. Chem.* **37**, 1470 (1965).

31. C. A. Bache and D. J. Lisk, *Anal. Chem.* **37**, 1477 (1965).

32. Y. Talmi and V. E. Norvall, *Anal. Chem.* **47**, 1510 (1975).

33. K. S. Brenner, *J. Chromatogr.* **167**, 365 (1978).

34. D. F. Hagen, J. S. Marhevka, and L. C. Haddad, *Spectrochim. Acta* **40B**, 335 (1985).

35. Z. Kewei, G. Qingyu, W. Guochen, and Y. Weile, *Spectrochim. Acta* **40B**, 349 (1985).

36. D. S. Sklarew, K. B. Olsen, and J. C. Evans, *Chromatographia* **27**, 44 (1989).

37. D. B. Hooker and J. DeZwaan, *J. Pharm. Biomed. Anal.* **7**, 1591 (1989).

38. D. B. Hooker and J. DeZwaan, *Anal Chem.* **61**, 2207 (1989).

39. D. B. Hooker and J. DeZwaan, *ACS Symp. Ser.* **479**, 132 (1992).

40. K. B. Olsen, D. S. Sklarew, and J. C. Evans, *Spectrochim. Acta* **40B**, 357 (1985).

41. P. L. Wylie, J. J. Sullivan, and B. D. Quimby, *J. High Resolut. Chromatogr.* **13**, 499 (1990).

42. N. Kovacic and T. L. Ramus, *J. Anal. At. Spectrom.* **7**, 999 (1992).

43. C. Bradley and J. W. Carnahan, *Anal. Chem.* **60**, 858 (1988).

44. S. Pedersen-Bjergaard and T. Greibrokk, *J. High Resolut. Chomatogr.* **15**, 677 (1992).

45. R. J. Skelton, P. B. Farnsworth, K. E. Markides, and M. L. Lee, *HRC & CC, J. High Resolut. Chromatogr. Chromatogr. Commun.* **11**, 75 (1988).

46. J. J. Kosman, *Am. Lab (Fairfield, Conn.)* **24**(8), 28T (1990).

47. J. L. Buteyn and J. J. Kosman, *J. Chromatogr. Sci.* **28**, 19 (1990).

48. S. M. Lee and P. L. Wylie, *J. Agric. Food Chem.* **39**, 2192 (1991).

49. W. S. Miles and B. D. Quimby, *Am. Lab. (Fairfield, Conn.)* **24**(8), 28F (1990).

50. P. Sandra, *Analusis* **20**(3), M32 (1992).

51. S. Rowland, R. Evens, L. Ebdon, and A. Rees, *Anal. Proc. (London)* **30**, 87 (1993).

52. J. A. Seeley, Y. Zeng, P. C. Uden, T. I. Eglinton, and I. Ericsson, *J. Anal. At. Spectrom.* **7**, 979 (1992).

53. D. Tanzer and K. G. Heumann, *Atmos. Environ.* **24A**(12), 3099 (1990).

54. D. Deruaz, A. Bannier, M. DeSage, and J. L. Brazier, *Anal. Lett.* **24**(9), 1531 (1991).

55. B. D. Quimby, P. C. Dryden, and J. J. Sullivan, *Anal. Chem.* **62**, 2509 (1990).

56. J. J. Sullivan, *Trends Anal. Chem.* **10**, 23 (1991).

57. P. C. Uden, *J. Chromatogr.* **313**, 3 (1984).

58. S. A. Estes, P. C. Uden, and R. M. Barnes, *Anal. Chem.* **54**, 2402 (1982).

59. R. Lobinski and F. C. Adams, *J. Anal. At. Spectrom.* **7**, 987 (1992).

60. H. Emteborg, D. C. Baxter, and W. Frech, *Analyst (London)* **118**, 1007 (1993).

61. R. Lobinski, W. M. R. Dirkx, M. Ceulemans, and F. C. Adams, *Anal. Chem.* **64**, 159 (1992)

62. T. M. Dowling and P. C. Uden, *J. Chromatogr.* **644**, 153 (1993).

63. R. W. Moshier and R. E. Sievers, *Gas Chromatography of Metal Chelates*, Pergamon, Oxford, 1965.

64. M. S. Black and R. E. Sievers, *Anal. Chem.* **48**, 1872 (1976).

65. P. C. Uden and T. Wang, *J. Anal. At. Spectrom.* **3**, 919 (1988).

66. P. C. Uden and Y. Zeng, *Chromatographia* **34**, 269 (1992).

67. Y. Zeng, J. A. Seeley, T. M. Dowling, P. C. Uden, and M. Y. Khuhawar, *J. High Resolut. Chromatogr.* **15**, 669 (1992).

68. S. A. Estes, P. C. Uden, M. D. Rausch, and R. M. Barnes, *J. High Resolut. Chromatogr.* **3**(9), 471 (1980).

69. P. C. Uden, R. M. Barnes, and F. P. DiSanzo, *Anal. Chem.* **50**, 852 (1978).

SELECTIVE CHROMATOGRAPHIC DETECTION BY PLASMA MASS SPECTROMETRY*

FRANCINE A. BYRDY AND JOSEPH A. CARUSO

Department of Chemistry, University of Cincinnati, Cincinnati, Ohio 45221-0172

7.1. INTRODUCTION

It has been well established that trace elements in the environment are of toxicological concern to many organisms including humans and animals. Concern is well warranted since toxic trace elements may be found in air, waters, soils, etc. and may pose carcinogenic, mutagenic, and/or teratogenic risks even at very low levels.

Since toxicity may occur at trace to ultratrace levels and the toxicity of an element depends on its chemical form and/or oxidation state, chemical speciation has become an important area of research. Speciation involves the determination and quantitation of the various chemical forms of a particular element. While solution nebulization is the type of sample introduction method most often used for plasma MS (mass spectrometry), chromatographic techniques such as high-performance liquid chromatography (HPLC), gas chromatography (GC), and supercritical fluid chromatography (SFC) provide the separation schemes necessary for investigators trying to obtain speciation information. Other researchers may only be interested in separating a variety of elements at trace levels (such as the rare earth elements) without the necessity for elemental speciation. Chromatographic techniques coupled with plasma source mass spectrometric detection may be useful for this purpose as well.

Owing to the excellent sensitivity and high selectivity of plasma source MS, it is a promising detection method for the analysis and speciation of

---

*Presented in part at the 1992 American Chemical Society National Meeting in Washington, DC, as Analytical Division paper No. 126.

Selective Detectors, edited by Robert E. Sievers. Chemical Analysis Series, Vol. 131.
ISBN 0-471-01343-9 © 1995 John Wiley & Sons, Inc.

trace elements. In comparison to atomic emission spectroscopy, detection limits are generally 2–3 orders of magnitude lower for plasma MS determinations. Additionally, isotope abundance information may be obtained and multielement studies may be done.

7.1.1. Speciation

The speciation of trace elements is currently a popular area of research in the field of elemental analysis. A driving force behind its popularity is the need for adequate toxicological sample assessment. Maintaining the integrity of the individual species to be analyzed is certainly of concern to researchers in this field. For example, analysis of trace elements in biological samples presents some problems and requires some precautions.[1] Commonly studied trace elements include arsenic, cadmium, chromium, gold, lead, iodine, iron, mercury, selenium, tellurium, tin, and the rare earth elements.

7.1.1.1. Arsenic

Occupational and environmental exposure to arsenic is of concern due to its known toxicity. Arsenic is used for various industrial and agricultural purposes such as in herbicides and pesticides. Exposure to arsenic is best assessed by speciation since the inorganic forms are more toxic than the organic species. Relative toxicities range from arsenite to arsenate, methylarsonic acid (monomethylarsinic acid, MMA), and dimethylarsinic acid (DMA), to the relatively nontoxic arsenobetaine.[2] Urine, blood, and hair are good indicators of human exposure.[3]

7.1.1.2. Cadmium

Cadmium in foodstuffs warrants concern due to the toxicity of this metal. Long-term effects of low-level exposure include emphysema and chronic obstructive pulmonary disease. Effects may also be seen on the cardiovascular and skeletal systems.[4] The degree to which cadmium is absorbed by the gastrointestinal tract depends on its chemical form, the effect that ligands (from the food and gut) have, and the action of digestive enzymes.[5]

7.1.1.3. Chromium

Chromium is used in many industrial processes such as steel alloying and leather tanning. Much interest has been placed on the toxicity of chromium compounds owing to their widespread use. While chromium is known to exist in all oxidation states from 0 to VI, Cr(III) and Cr(VI) are the forms most commonly found.[6] There is a "chromium paradox"[7] in that Cr(III) is an essential trace nutrient whereas Cr(VI) is carcinogenic and mutagenic: Cr(III) is present in most soils but is less mobile in the environment than is Cr(VI); Cr(VI) compounds are less frequently found in soils but are very toxic to aquatic plants and animal life and are frequently used in algaecides.[6,8]

7.1.1.4 Gold

Gold-containing drugs have been used for more than 60 years in the treatment of rheumatoid arthritis. These drugs include Myochrysine [sodium gold(I) thiomalate], Ridaura [auranofin or (triethylphosphine)gold(I) tetraacetyl-thioglucose], and Solganal [gold(I) thioglucose].[9] There are frequent side effects associated with Myochrysine, and treatment must be stopped in roughly 25% of the cases owing to the severity of these effects. In 2% of the cases undergoing auranofin treatment, usage must be discontinued. Identification and quantification of the drugs and their metabolites will certainly provide insight into the pharmacological properties and subsequent actions of these drugs.[10]

7.1.1.5. Lead

It is well known that lead has adverse effects on health. Even trace levels pose a threat, especially to children, since they absorb lead more efficiently than adults.[11] The harmful species include inorganic lead, which is found in older house paints, tetraalkyllead, once used in the United States as an antiknock agent for gasoline, tri- and dialkyllead found in waters, soils, fish, etc., and other organolead compounds that are thought to be derivatives of natural biotic and abiotic methylation.[12] Alkyllead compounds are more toxic than inorganic lead, and owing to the cumulative nature of lead, speciation at trace levels is of great importance.[11]

7.1.1.6. Iodine

Radioiodine is absorbed rapidly into the bloodstream and concentrates in the thyroid. The toxicity of ^{131}I is of particular interest owing to its widespread use in nuclear medicine and in fallout after testing of nuclear weapons. Increased levels of the radioactive ^{131}I results in almost complete destruction of the thyroid and markedly decreased thyroid hormone production.[4]

7.1.1.7. Iron

Fe(II) and Fe(III) are the iron oxidation states most commonly found. Accordingly, the speciation of these two states is reported most. Simultaneously determining these species is of importance to the understanding of the env"iron"mental redox processes occurring in biological systems (e.g., in porphyrins in hemoglobin, myoglobin, cytochromes, peroxidases, and catalases). It is also important in the assessment of water quality, since iron affects various chemical processes in waters.[6]

7.1.1.8. Mercury

There are a wide variety of effects caused by the different biochemical forms of mercury. Toxicologically, there are three basic forms of interest: elemental mercury, inorganic mercury, and organomercury compounds. In terms of toxicity, methylmercury is the most important form since it has a selective effect on cerebellum and visual cortex.[4]

7.1.1.9. Selenium

Selenium is found in igneous rocks and copper ores, and is used as an additive in ceramics, rubber, and photographic processes. Like chromium, it is both toxic and essential. Like lead, selenium has a cumulative toxic nature. In high concentrations it poses health hazards, whereas it is an essential trace nutrient at very low levels ($\mu g \ kg^{-1}$). The major reaction chemistry of selenium rests on selenate (SeO_4^{2-}), selenite (SeO_3^{2-}) and selenide (Se^{2-}). Common oxidation states of selenium include II, IV, and VI, with the last-named species being far more stable than Se(IV).[6]

7.1.1.10. Tellurium

Tellurium is a highly toxic metal widely used in industry. It is used to improve the acid resistance of lead in batteries and is employed as an alloy constituent in metallurgy. Tellurium-doped single crystals used in the semiconductor industry contain ultratrace levels of tellurium. While oxidation states II, IV, and VI are the most common, Te(IV) is the most stable.[6]

7.1.1.11. Tin

Organotin componds are used as catalysts, biocides, and polymer stabilizers. Fungicides, antifouling paints, and wood preservatives commonly contain trialkyltin compounds. These compounds are toxic, and their extensive release into the environment is of concern.[13] Additionally, the two stable inorganic species, Sn(II) and Sn(IV), are important in assessing plating bath performance.[6]

7.1.2. Plasma MS

Inductively coupled plasmas (ICP) and microwave-induced plasmas (MIP) are two commonly used plasma ion sources. The ICP is the most frequently used and is formed by passing radio-frequency (RF) energy (usually at 27.1 MHz) through a load coil to form an oscillating magnetic field. A plasma forms when a gas, usually argon, is "seeded" by a spark with electrons and thus becomes conductive. Energies sufficient to ionize gaseous atoms in the magnetic field are reached. The plasma will become self-sustaining because of collisions with gaseous atoms and ohmic heating, causing further ionization. The ICP is formed in a quartz torch where the sample will travel to be vaporized, desolvated, and ionized. Sample introduction is usually accomplished by means of a nebulizer, which creates an aerosol from solution by passage through a small orifice with high-velocity gas; also used is a spray chamber, whose purpose is to separate out the large droplets of sample that go to waste.[14] Table 7.1 shows some typical ICP/MS operating conditions.

Table 7.1. Typical ICP/MS Operating Conditions

Inductively coupled plasma:	
Incident RF power	1250 W
Reflected RF power	< 10 W
Coolant argon flow rate	15 L min^{-1}
Auxiliary argon flow rate	< 0.5 l min^{-1}
Nebulizer argon flow rate	0.700 L min^{-1}
Sampling position	On center; 10–15 mm from the load coil
Vacuum system:	
Expansion pressure	1.5–2.5 mbar
Primary stage	< 10^{-4}
Analyzer stage	5 × 10^{-6}
Interface:	
Sampling cone	Nickel, 1 mm orifice
Skimmer cone	Nickel, 1 mm orifice
Sampler–skimmer separation	6 mm
Sample introduction:	
Nebulizer	Concentric
Spray chamber	Double-pass, cooled to 5 °C
Sample flow rate	1 mL min^{-1}

Source: Reprinted from Heitkemper and Caruso,[17] with permission of Elsevier Science Publishers B.V.

Mass spectrometric detection coupled with an argon plasma ion source has become important for trace elemental research owing to its excellent sensitivity and selectivity and its potential for isotope determinations. ICP/MS provides the researcher with detection limits necessary for environmental and biological studies. In addition, elemental MS spectra are relatively simple compared to optical spectra. Disadvantages include high instrumental cost and formation of interfering polyatomic species from matrix elements and atmospheric and plasma gases.[15] Potential interferences are discussed in detail elsewhere[16].

A microwave-induced plasma is an alternative ion source. A quartz discharge tube is generally placed in the middle of a resonant cavity that is connected to a microwave generator by a coaxial cable. The plasma gas is usually helium. Unlike ICP/MS, there is no commercially available MIP/MS on the U.S. market today, most likely because of concerns as to its perceived operational difficulties. The major disadvantage with helium plasmas is their intolerance to aerosols at low powers. When an MIP is used, however, there are two potential advantages over an argon ICP. First, argon isotope interferences should be eliminated, and the interferences arising from the combination of plasma gas and matrix elements should occur at lower m/z (mass-to-charge) values. Second, more energetic plasmas are possible with helium plasmas in comparison to the use of argon and nitrogen gases for plasma formation. Argon plasmas do not ionize higher ionization potential nonmetals such as the halogens, sulfur, and phosphorus as efficiently as a helium plasma does.[14,17,18]

7.2. HIGH-PERFORMANCE LIQUID CHROMATOGRAPHY

HPLC is the most common chromatographic technique used with ICP/MS detection. An extensive list of references starting with the most recently published in English is provided for the interested reader.[2,3,5,9–13,19–66] One reason for this popularity is the abundance of HPLC separations in the literature from which researchers can draw. Another reason is the simplicity of the interface. All that is required is tubing to connect the outlet of the analytical column with the inlet of the nebulizer. Finally, commonly used flow rates in HPLC (on the order of 0.5–2.0 mL/min) are compatible with the flow rates used in most commercial ICP instruments. Figure 7.1 shows the instrumental setup for HPLC/ICP/MS.

Problem areas for HPLC/ICP/MS analyses include possible clogging of the nebulizer, torch, and sampler and skimmer orifices when mobile phases containing high total dissolved solids (greater than 0.2%) are employed. Also, the high concentration of organics generally used in HPLC mobile phases

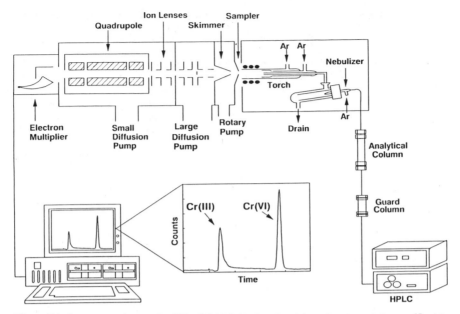

Figure 7.1. Instrumental setup for HPLC/ICP/MS. Reprinted from Byrdy and Caruso,[67] with permission of the Public Health Service, U.S. Dept. of Health & Human Services.

may result in sensitivity losses and plasma instability. Various researchers have found ways to circumvent these problems, and these will be discussed within the context of this section.

The analysis of trace metals by HPLC/ICP/MS may be divided into three main categories depending on the mode of separation used: (1) reversed-phase liquid chromatography (RPLC), which includes variations such as ion-pairing and micellar liquid chromatography; (2) ion-exchange chromatography (IEC), which is sometimes referred to as ion chromatography; and (3) size-exclusion chromatography (SEC). Table 7.2 shows detection limits for a select number of elements investigated by various HPLC/ICP/MS modes.

7.2.1. Reversed-Phase Liquid Chromatography

There have been a number of reports of separations achieved by RPLC.[10,19,23,26,29,31,34,45,53,57,60,61] It is the most popular mode of separation and is accomplished by partitioning the analyte of interest between a nonpolar stationary phase and a polar mobile phase. Mobile phases may include mixtures of methanol, tetrahydrofuran, or acetonitrile, and water.

Table 7.2. Detection Limits for Select Elements Investigated by HPLC/ICP/MS[a]

Element	HPLC Mode[b]					Absolute Detection Limit (pg)	References
	RP	IP	MLC	IEC	SEC		
As		×				0.2–1000	36,[c] 66
				×		20–1000	2, 3, 20, 33, 49, 50
	×					300	60
Cd		×				300	66
				×		60	44
					×	0.5	5, 24,[d] 47
Cr		×				1000	66
				×		1300–1900	28, 48
Au		×				300–2500	9, 10, 66
					×	NR	56
			×			NR	56
I	×					0.9–130	26, 45[d]
				×		1.5	64
Pb				×		40	44
	×					25–87	11
		×				0.2–3900	11, 12, 35,[c] 66
					×	0.5	24[d]
Hg		×				6–18	35[c]
	×					60–2000	53, 61
Sn		×				1.5–1000	13, 21, 23, 36[c], 46, 58
				×		400–1000	58
			×			27–130	59

[a]No attempt has been made to normalize the detection limits for the calculation method or compound type.
[b]Abbreviations: NR = not reported; RP = reversed-phase; IP = ion-pairing; MLC = micellar liquid chromatography; IEC = ion-exchange chromatography; SEC = size-exclusion chromatography.
[c]Using direct injection nebulization.
[d]MIP/MS detection.

7.2.1.1. Applications

Bushee used RPLC and a Waters PicoTag C_{18} column to speciate mercury, and the method was applied to the determination of methylmercury in an NBS RM-50 albacore tuna sample $(873 \pm 60 \, ng \, mL^{-1})$.[61] Additionally, thimerosal, a mercury-containing preservative found in contact lens solutions, was determined. Detection limits were lowered by using postcolumn mercury cold-vapor generation. Also, flow injection was employed for total mercury determinations due to "mercury memory" when the direct ICP/MS mode was used.

Another study by Bushee et al. involved a long-term stability analysis of thimerosal in injectable biological products.[53] The rubber stoppers used to contain the samples were found to contain thimerosal, thereby representing mercury absorption by the stoppers. Samples containing thimerosal as a preservative (vaccines, toxoids, and diluents) were analyzed. Samples were also investigated for the presence of decomposition products such as methyl-mercury chloride, dimethylmercury and mercury(II) chloride. Besides an unknown minor mercury compound in one of the samples, none of the listed decomposition products was found.

Beauchemin et al. used RPLC/ICP/MS, thin-layer chromatography, elec-tron impact MS, and graphite furnace atomic absorption spectrometry to identify and quantitate arsenic species in a dogfish muscle reference material (DORM-1).[60] The major constituent of the reference material was found to be arsenobetaine (15.7 ± 0.8 µg of As per gram of DORM-1, which is roughly 84% of the total arsenic content). For arsenobetaine, the absolute detection limit of arsenic by RPLC/ICP/MS was 0.3 ng. A methanol/chloroform extraction step was used to isolate this species.

A recent publication by Takatera and Watanabe describes the speciation of iodo amino acids by RPLC/ICP/MS.[26] Iodide ion and five iodo amino acids commonly found in thyroglobulin (monoiodotyrosine, diiodotyrosine, 3,3',5- and 3,3'5'-triiodothyronine, and thyroxine) were separated. Absolute detection limits ranged from 35 to 130 pg as iodine. These detection limits are approximately 1 order of magnitude lower than those achieved by conventional stable isotope iodine methods. Additionally, the iodine-containing compounds in an enzymatic digest of bovine thyroglobulin were determined.

Heitkemper and co-workers remain, as of this writing, the only researchers to use RPLC with He-MIP/MS detection.[45] For the analysis of halogenated organic compounds, an Alltech Econosphere C_{18} column was used with a mobile phase consisting of 70% v/v HPLC-grade methanol. Absolute detection limits for brominated compounds (50 pg Br), iodinated compounds (1 pg I), and chlorinated compounds (10 ng Cl) were found.

Metalloporphyrins (cobalt and zinc protoporphyrins and hemin) have been separated by reversed-phase HPLC.[19] Owing to the elemental selectivity of ICP/MS, the porphyrins were separated from interfering metabolites and other metalloporphyrins without the need for derivatization. Excellent detec-tion limits were obtained for cobalt and zinc protoporphyrins, and the accuracy of the method was assessed by extraction and determination of zinc protoporphyrin in whole blood from a patient who was poisoned with lead.

7.2.2. Ion-Pairing Chromatography

The addition of a counterion to the mobile phase constitutes reversed-phase ion-pairing chromatography (IP). The primary advantage IP has over regular

RPLC is that while ionic species are not retained by regular RPLC, IP facilitates the analysis of both ions and molecular species.[68] Commonly used ion-pairing reagents include tetrabutylammonium phosphate (PIC-A) for acidic samples and sodium pentanesulfonate (PIC-B5) for basic samples. A disadvantage of the technique is short column lifetimes when corrosive ionic solutions are used.[69] Despite this drawback, many reports have been published utilizing this technique.[9–13,21,22,25,35,36,50,58,63,66]

7.2.2.1. Applications

The first published work employing ion-pair RPLC with ICP/MS detection was also the first reported use of plasma MS as an HPLC detector.[66] Thompson and Houk evaluated the feasibility of using ICP/MS as a multi-element detector for flow injection analysis (FIA) and ion-pair RPLC. An Alltech Econosphere C_{18} column was used with PIC-B5 and PIC-A ion-pairing reagents. The speciation of As and Se was achieved with sub-ng detection limits. Additionally, the ability to measure isotope ratios and multi-element detection were demonstrated. Jiang and Houk continued work on RPLC/ICP/MS,[63] and another paper describes the separation of anionic compounds containing phosphorus or sulfur. Tetraalkylammonium salts were used as ion-pairing reagents.

Recently, in our laboratories, Kumar et al. have investigated the effect that inorganic tin chloride has on the separation of trimethyl, tributyl, and triphenyltin chlorides.[13] Two different columns were investigated: a Spherisorb ODS-2 C_{18} (silica) column, and a PRP-1 (polymeric) column. Use of the former column was previously reported by Suyani and co-workers.[58] A relatively high background was observed and was likely a result of the accumulation of tin compounds on the column from the equilibration of tin between the column packing material and the ion-pair reagent.

Inorganic tin was found to be retained strongly on both columns. It was held to a greater extent, however, on the silica column. As a result, the use of the PRP-1 column was favored over the silica-based column. The ion-pairing agent used in Kumar and colleagues' study was PIC-B5. Mobile phase at a pH of 6 was found to be optimal and analysis time was roughly 6 min in comparison to 13 min when the ODS-2 column was used. Figure 7.2 shows the effect of pH on the separation of tin. The detection limits were lowered for the organotin compounds from the nanogram level obtained by Suyani et al. to the picogram level. The method was applied to tributyltin chloride (TBT–Cl), and triphenyltin chloride (TPT–Cl), which were extracted from a certified fish tissue reference material and grocery store tuna fish.

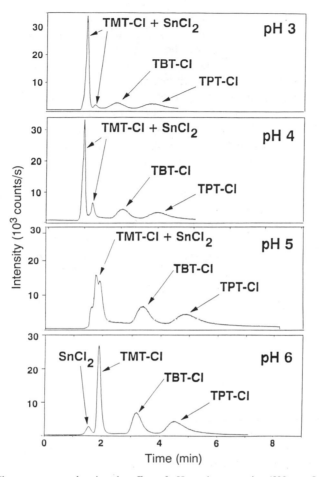

Figure 7.2. Chromatograms showing the effect of pH on tin separation (500 ng mL^{-1} SnCl$_2$ + 100 ng mL^{-1} of organotin compounds) on PRP-1 column with mobile phase consisting of 94% methanol, 5% water, 1% acetate (e.g., for pH = 6; 0.046 acetic acid and 0.012 M ammonium acetate were used), and 0.004 M sodium pentane sulfonate; flow rate = 1 mL/min. Reprinted from Kumar et al.[13] with permission of Elsevier Science Publishers B.V.

A follow-up study by Kumar and co-workers involved supercritical fluid extraction of organotin from biological samples (e.g., fish tissue and tunafish).[21] The procedure described above was used for HPLC/ICP/MS analyses. The amount of sample required for analysis was reduced to 0.14 g from the 2.5 to 5 g previously reported using solvent extraction.[13]

Others at the University of Cincinnati have used ion-pairing RPLC to study gold-based antiarthritis drugs and their metabolites utilizing ICP/MS detection.[10] Zhao et al. used a Spherisorb ODS-2 C_{18} column for method development, while a Burdick and Jackson OD5 C_{18} column was used for biological sample analyses. For all experiments, the mobile phase consisted of methanol/water, buffer, and a variety of ion-pairing agents. It was found that tetra-n-butylammonium chloride (TBA) did not alter neutral molecule retention, so it was chosen as the ion pairing agent.

Other separation conditions such as pH, methanol content, ionic strength, and the concentration of TBA were also studied. The absolute detection limit for auranofin was 0.3 ng. The method was applied to the analysis of urine from arthritis patients who were undergoing auranofin, Myochrysin, or Solganal therapy. In the urine of several of the patients on gold-based therapy, dicyanogold(I) anion was found. This discovery led to another paper on the analysis of dicyanogold(I) in the blood and urine of patients receiving sodium gold(I) thiomalate for treatment of arthritis.[9] The optimal chromatographic conditions were used from the previous study.[10]

The studies of Zhao, and co-workers were the first to demonstrate the presence of dicyanogold(I) in patient fluids. Also, their work supported the idea that the metabolites of gold drugs are the active species rather than the administered drug. The most recent work of this group involved the determination of cisplatin and possible metabolites for the specific detection of platinum.[25] A Burdick and Jackson OD5 C_{18} column was utilized with sodium dodecyl sulfate (SDS) or 1-heptanesulfonate as ion-pairing agents. The detection limit was 0.1 ng for platinum. When heptanesulfonate was used as the counterion, all platinum complexes of interest were resolved in a faster separation than when SDS was used. The goal was to elucidate cisplatin metabolism. The identity of metabolite peaks for future work was sought, and hydrolysis studies of cisplatin were done as well as studies of the reaction of cisplatin with biologically important thiols.

7.2.3. Micellar Liquid Chromatography

The addition of detergents or surfactants as counterions is termed micellar liquid chromatography. SDS is the most commonly used surfactant. It is anionic in nature, while other common surfactants, e.g., cetyltrimethylammonium bromide (CTAB) and polyoxyethylene(23) lauryl ether (Brij-35), are cationic and nonionic, respectively. Micelle solutions have the ability to solubilize hydrophobic compounds in aqueous solution. Through the minimization of organic solvent usage, other advantages such as low cost, low toxicity, and greater plasma compatibility are obtained.[59]

7.2.3.1. *Applications*

To date, the only use of a micellar mobile phase for separation with plasma MS detection was reported by Suyani and co-workers.[59] A C_{18} stationary phase and a 0.1 M SDS micellar mobile phase with 3% v/v acetic acid and 3% v/v propanol were used to speciate alkyltin compounds. Tin detection limits for trimethyltin chloride (TMT–Cl), triethyltin bromide (TET–Br), and tripropyltin chloride (TPT–Cl) were 27, 51, and 111 pg, respectively. In addition, a 0.02 M SDS mobile phase with 0.010 M KF, 3% v/v acetic acid, and 3% v/v propanol was utilized in another separation scheme for organotin compounds. Tin detection limits were 46, 26, and 126 pg, respectively, for monomethyltin trichloride (MMT–TCl), dimethyltin dichloride (DMT–DCl), and TMT–Cl. Preliminary separation of all five organotin compounds was done by gradient elution. The chromatogram obtained is given in Figure 7.3.

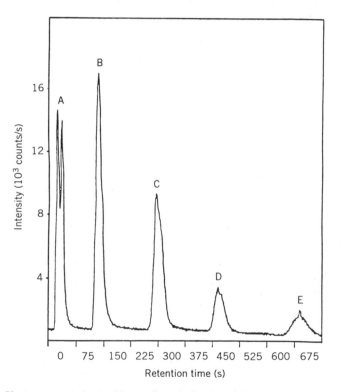

Figure 7.3. Chromatogram obtained by gradient elution from 0.02 to 0.1 M SDS for 2 min, and constant at 0.1 M SDS after 2 min: (A) MMT–TCl; (B) DMT–DCl; (C) TMT–Cl; (D) TET–Br; (E) TPT–Cl (8 ng tin each). Reprinted from Suyani et al.[59] with permission of the Society for Applied Spectroscopy.

7.2.4. Ion-Exchange Chromatography

The majority of elemental speciation separations have been performed by IEC.[2,3,11,20,24,28,31,33,41,42,44,46,48–50,54,56,58,64] A reason for its popularity stems from its potential to separate both ionic and nonionic species. This includes complex ions and neutrals. The key factor in separation is pH. Selectivity, sample retention, and the extent of ionization all depend upon pH. The mobile phase is generally a buffered solution of an aqueous salt with a small amount of methanol or acetonitrile, and the resolution of non-polar compounds is possible by ligand-exchange reactions or ionic complex formation prior to separation. Typically, the term *ion chromatography* (IC) denotes the use of a second column to suppress mobile phase ionic strength in order to perform conductimetric detection of the analyte ions of interest; it has, however, often been used interchangeably with IEC. Separations achieved by IC are IEC separations.[69] Gjerde and Mehra have written an excellent review of ion chromatographic speciation techniques for trace metals analysis.[6]

7.2.4.1. Applications

Several papers from our research laboratory have discussed the speciation of arsenic by anion-exchange HPLC.[2,3,49] Heitkemper et al. used a weak anion-exchange column with a 5 μm Adsorbosphere–NH$_2$ packing from Alltech to separate four arsenic compounds commonly found in urine: arsenite, arsenate, DMA, and MMA.[2] For arsenic in an aqueous medium, absolute detection limits were in the range of 20 to 91 pg. For a freeze-dried urine sample, detection limits were 36–96 pg. An interference from ArCl$^+$ complicated the determination of arsenite and was dealt with by Sheppard et al., who used ion chromatography to speciate arsenic species found in urine.[49] (This work will be discussed in detail below in Section 7.2.6.1.5.) The most recent work of Sheppard and Caruso involved the separation and detection of four arsenic compounds in freeze-dried urine, club soda, and wine.[3] Figure 7.4 shows the separation of four arsenic species and chloride. An improvement in sensitivity was possible by using a mixed-gas helium–argon plasma. Unfortunately, the ArCl$^+$ interference increased when the mixed-gas plasma was used.

A paper by Honoré Hansen et al. describes the separation of seven arsenic compounds by HPLC with both hydrogen–argon flame atomic absorption spectrometry (FAAS) and ICP/MS detection.[33] Both anionic- and cationic-exchange HPLC modes were employed for this study. The ICP/MS system was used for the comparison of results obtained for arsenic in soil by

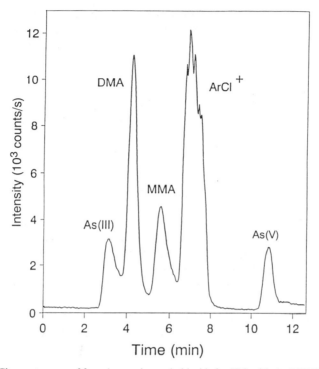

Figure 7.4. Chromatogram of four As species and chloride by IEC with Ar ICP/MS detection. All As species were at concentrations of $100 \, \mu g \, dm^{-3}$ As in $1 + 4$ urine. Mobile phase: 50 mmol dm^{-3} carbonate/hydrogen carbonate buffer at pH 7.5 using a gradient program. Sample loop = $100 \, mm^3$. Monitoring at m/z 75. Reprinted from Sheppard and Caruso[3] with permission of the Royal Society of Chemistry.

HPLC/FAAS. The authors suggest that the HPLC/FAAS system is applicable to only certain practical arsenic speciation problems.

Roehl and Alforque compared ion chromatography (IC) ICP/MS with colorimetry in the determination of hexavalent chromium.[48] Similar detection limits and dynamic ranges were obtained for both detection schemes. By means of IC/ICP/MS, oxyanions of other metals (As, Se, V, Mo, and W) were separated. In a subsequent paper,[42] Riviello et al. described the use of IC to enhance the performance of ICP/AES and ICP/MS. The authors discussed how coupling these techniques can solve various problems that arise in the determination of trace elements in biological matrices. Within their discussion, chelation concentration is explained and its usage discussed in detail.

Other workers have used IC with ICP/MS detection. Suzuki et al. have reported the analysis of inorganic impurities in methamphetamine.[64] Fourteen trace rare elements were separated by Kawabata and co-workers.[41] Finally, Klinkenberg et al. determined tellurium compounds by IC/ICP/MS.[46]

7.2.5. Size-Exclusion Chromatography

The packing material in SEC columns contains pores capable of distinguishing between analytes. Separation by SEC, formerly known as gel-permeation chromatography, depends on the molecular size of the compound of interest. There are only a few reports of SEC coupled to ICP/MS.[5,24,31,34,38,39,47,56,62,65]

7.2.5.1. Applications

Dean et al. have investigated trace metals in metalloprotein species with ICP/MS detection.[65] A standard of metallothionein, from horse kidney, and ferritin were separated and the cadmium content determined. A later publication by Crews and co-workers discussed the speciation of cadmium in pig kidney.[5] Results indicated that the majority of soluble cadmium in the sample was associated with a metallothionein-like protein. The latter protein withstands cooking and simulated gastrointestinal digestion *in vitro*. The relative molecular mass ranged from 1.2×10^6 to 6×10^3 in three fractions of uncooked pig kidney. Upon enzymatic treatment or cooking, only the molecular mass fraction was found.

Mason et al. have determined cadmium, zinc, and copper in metallothionein,[47] and Gercken and Barnes have determined lead and other trace elements in blood components by SEC/ICP/MS.[39] Figure 7.5 shows the chromatogram of lead, copper, and zinc distribution in rat blood serum. Lead was detected in at least four molecular weight fractions, as denoted on the chromatogram. The detection limit for lead in protein fractions was $0.15\,\mu g\,L^{-1}$.

7.2.6. Comparison Studies

Some researchers have utilized more than one mode of HPLC to achieve an optimal separation. For example, Beauchemin et al. used ion-pairing and ion-exchange to determine arsenic species.[50] As(III), As(V), MMA, DMA, and arsenobetaine (AsB) were separated with detection limits between 50 and 300 pg as arsenic. The same extraction procedure used previously[60] was employed to extract arsenic from a dogfish muscle reference material (DORM-1). Anion-pairing, cation-pairing, and anion-exchange modes were all investigated. Results indicated that anion-exchange was most tolerant

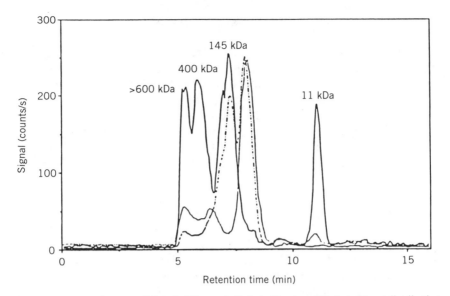

Figure 7.5. Chromatogram of Pb (m/z 208, ——), Cu (m/z 63,– –), and Zn (m/z 64, ·····) distribution in rat blood serum; ^{63}Cu count rate has been divided by 5, and ^{64}Zn by 10. Evaluated molecular weights of lead fractions are labeled. Reprinted with permission from Gercken and Barnes.[39] Copyright 1991, American Chemical Society.

because of the high buffering capacity of the mobile phase, cation-pairing was found suitable for the determination of DMA and AsB; anion-pairing was most sensitive to changes in the sample matrix.

Matz et al. used SEC and IEC to analyze blood samples from rheumatoid arthritis patients on gold drug therapy.[56] Copper, gold, and zinc were investigated. Examples of SEC, RPLC, and weak anion-exchange chromatography are given in a chapter by Elder et al.[31] Samples studied were once again from rheumatoid arthritis patients.

Suyani and co-workers looked at ion-exchange and ion-pairing chromatography with both ICP/AES and ICP/MS detection schemes for the speciation of organotin compounds.[58] Linear response was better (over 3 orders of magnitude) with IEC than with ion-pairing chromatography (over 2 orders of magnitude). Al-Rashdan et al. speciated lead by HPLC/ICP/AES and HPLC/ICP/MS.[11] Reversed-phase, ion-pairing, and ion-exchange HPLC modes were studied. Destabilization of the plasma was found using gradient elution and ICP/MS detection. This destabilization was addressed in a subsequent paper.[12]

7.2.6.1.   Special Studies

7.2.6.1.1.   On-Line Preconcentration of Trace Metals. Preconcentration is often necessary when samples contain ultratrace concentrations of elements which are below the detection capability of the ICP/MS. For example, levels of Cd and In may be below the detection capability in a number of typical environmental samples.[70] Boomer and co-workers used a cation-exchange column (Dionex CS5) for trace metal preconcentration employing a high-performance ion chromatograph (HPIC).[44] Copper, cadmium, mercury, and lead were studied, with the optimum eluent strength for the determination of Cu, Cd, and Pb being 5% v/v nitric acid. A flow rate of 3.5 mL/min was used.

7.2.6.1.2.   Gradient Elution. All but two of the chromatographic separations discussed thus far[12,59] were accomplised by isocratic elution. This means that the composition of the mobile phase was not changed during the course of a chromatographic run. When, however, the sample contains analytes which have a wide range of partition ratios, gradient elution may be performed by altering the mobile phase composition during a run. This mode of operation is very powerful since analysis times may be reduced, and improvements in both peak shape and limits of detection may be achieved. A disadvantage of gradient elution is the necessity to reequilibrate the column prior to each run. This process is often very time- and solvent-consuming. Ion-pair reversed-phase chromatography with gradient elution was used by Al-Rashdan et al. to speciate triethyllead chloride (TEL-Cl), tetraethyllead (TtEL), triphenyllead chloride (TPhL-Cl), and inorganic lead (Pb(II)).[12] An Alltech Nucleosil C_{18} column was used with a PIC-B5 ion-pair reagent. Lead detection limits were 0.37 ng for Pb(II), 0.14 ng for TEL, 0.17 ng for TPhL, and 3.9 ng for TtEL. The viability of the method was verified by using water and fuel reference materials.

7.2.6.1.3.   Multielement Time-Resolved Analysis. SEC was used by Owen and coworkers to separate a known mixture of proteins.[34] Additionally, reversed-phase chromatography was employed for the separation of Zn-containing species in chicken that was both intrinsically and extrinsically labeled with stable Zn isotopes. Time-resolved analysis (TRA) software was used to reduce the number of chromatographic runs by permitting rapid data acquisition. By use of TRA software, continuous multi-isotope or multielement monitoring of HPLC eluent may be achieved. The entire mass range from 4 to 255 u was scanned in 100 ms and data was acquired in either the scanning or peak-jumping modes for a total acquisition time depending on the length of the chromatographic run. Unfortunately, data handling was found to be cumbersome.

7.2.6.1.4. Direct Injection Nebulization. Pneumatic nebulization through a concentric or cross-flow nebulizer with a spray chamber is the most common sample introduction method for ICP/MS. Because of several shortcomings, including the inefficiency of the system (98% of the sample goes to drain), ultrasonic and direct injection nebulizers (DIN) have been constructed. The DIN is a microconcentric pneumatic nebulizer inside the ICP torch; its use serves to reduce wash-out times for memory-prone elements such as Hg, I, and B. The dead volume is low and, in combination with the lack of a spray chamber, postcolumn band broadening is lessened. The use of an HPLC microcolumn is facilitated with 30–100 µL/min flow rates.[35] The DIN provides nearly 100% sample throughput, but it lacks a desolvation step. Also, specialized torches are required and detection limits are only 1–2 orders of magnitude better than those obtained with conventional nebulizers.

Shum et al. used a DIN to speciate various mercury and lead cationic species using ion-pairing agents selected based on their chromatographic performance and detection sensitivity.[35] Detection limits for lead as Pb^{2+}, $(Me)_3Pb^+$, and $(Et)_3Pb^+$ were roughly 0.2 pg. Detection limits for mercury as Hg^{2+}, $MeHg^+$, $EtHg^+$, and $PhHg^+$ were about 7 pg. The inorganic lead and mercury compounds were monitored directly in human urine. Organolead and organomercury ions were seen only when spiked into the sample.

A new version of the DIN was used by Shum et al. to separate charged species of arsenic and tin.[36] Ion-pairing chromatography was used on a microscale (column i.d. of 1 mm). One important requirement for microscale HPLC is low extracolumn band broadening, and the low dead volume (< 1 µL) of the DIN afforded this requirement. Detection limits were 0.2–0.6 pg for arsenic and 8–10 pg for tin. These researchers found improved absolute detection limits and chromatographic resolution with the use of the DIN. Additionally, this nebulizer has shown potential for use with gradient elution, and its use has allowed the direct injection of HPLC eluents containing large amounts of organic modifier (up to 85% MeOH).

More recently, Shum et al. have discussed the separation of two Se species by anion-exchange HPLC and metalloproteins in human serum (NIST SRM 909a-1) by SEC.[24] Both methods employed a DIN with packed microcolumns. The metalloproteins were directly measured in human serum without extraction or preconcentration. Finally, the ability to use chromatographic peak areas for isotope ratio measurements on Se species was demonstrated. This suggests the potential to measure isotope ratios for elements in "tough matrices" (e.g., serum, urine, and seawater).

7.2.6.1.5. Argon Chloride Interference Elimination. As discussed earlier in the text, there is an $^{40}Ar^{35}Cl^+$ interference found at the same mass as arsenic. This would greatly disrupt any arsenic speciation studies of human urine (a primary source of As toxicological information), which contains roughly 0.15 M sodium chloride. Sheppard and co-workers have devised a way to overcome this problem by using ion chromatography to separate chloride from the arsenic species of interest and by using a sample dilution factor to avoid column overload by Cl^-.[49] Unfortunately, higher detection limits in urine were obtained owing to the 20-fold dilution factor, and DMA and MMA were not separated completely. Detection limits were 3.4 ppb for As(III), 4.2 ppb for As(V) and 7 ppb for DMA. The mobile phases used in this study employed phthalic acid. Ammonia solution was used to make pH adjustments.

7.2.7. Future Developments

Speciation and multielement studies employing HPLC/ICP/MS are numerous. Since 1986, many techniques have been developed using various HPLC modes for a host of applications with ICP/MS detection. Crews and co-workers, however, have pointed out that there are "potential pitfalls" in the use of HPLC for trace inorganic speciation studies.[5] The behavior of the analyte of interest on the chosen column must be clearly investigated, since the speciation of the analyte may be altered during the analysis itself. Also, anomalous peaks may arise from retention of trace inorganic ions from equipment, samples, etc. In conclusion, care must be taken to maintain the integrity of the analyte. Moreover, a wider range of real samples needs to be made amenable to separation by HPLC techniques. Speciation studies in our laboratory focusing on tin, arsenic, and lead have led us into current work on chromium and vanadium. Certainly, the development of HPLC techniques to analyze species of toxicological interest at sub-ppb levels will continue to be a major area of research.

7.3. GAS CHROMATOGRAPHY

Unlike solution nebulization, gaseous sample introduction theoretically provides nearly 100% analyte transport efficiency to the plasma.[14] Since the need for desolvation and vaporization are removed, additional plasma energy may be available for ionization.[71] Helium MIP/MS is the most common plasma interface between GC and MS.[72–79] One reason is that helium plasmas ionize nonmetals such as the halogens more efficiently than does the ICP.

Few GC/ICP/MS studies have been done.[32,80–83] These studies have been limited to organometallics with methyl, ethyl, propyl, or butyl groups so as to obtain the volatility required for GC analysis. In order to be amenable

Table 7.3. Detection Limits for Some Elements Investigated by GC/Plasma MS[a]

Element	GC Mode[b]	Absolute Detection Limit (pg)	References
Br	LP/MIP	3.5	74
	AP/MIP	0.92–1.1	76, 77
	AP/ICP	100[c]	83
Cl	LP/MIP	22–36	72, 74, 77
	AP/MIP	9.2–21	76
	AP/ICP	13,000–18,000	80
	AP/ICP	40,000[c]	83
I	LP/MIP	0.1	74
	AP/MIP	1.5	76, 77
	AP/ICP	1.0[c]	83
Pb	AP/ICP	0.7[c]	81
P	LP/MIP	86–790	72, 73
	AP/ICP	1.0[c]	83
S	LP/MIP	97–510	72, 73
	AP/ICP	3,000[c]	83
Sn	AP/MIP	0.09–4	75
	AP/ICP	7.4–21	80
	AP/ICP	3.0–6.5[c]	82

[a] No attempt has been made to normalize the detection limits for the calculation method or compound type.
[b] Abbreviations: LP = low pressure; AP = atmospheric pressure.
[c] Picograms per second.

to separation by GC, compounds must be volatile and thermally stable. Compounds that are not volatile or thermally stable may be derivatized to meet these requirements.[71] Table 7.3 lists some of the elements investigated by various GC/plasma MS schemes.

7.3.1. GC/MIP/MS

In 1965, McCormack and colleagues were the first to interface a gas chromatograph to an MIP[84] and a review of subsequent GC/plasma AES determinations has been given.[85] In the early part of the 1980s, Douglas and co-workers described an argon MIP/MS system.[86] Since then, the majority of GC/plasma MS applications have utilized MIP/MS detection for the analysis of halogenated, nonmetallic, and metallic compounds. Helium is the carrier gas typically used in GC/plasma MS since plasma energy may be lowered when other gases such as nitrogen or hydrogen are used.[71] The MIP has lower gas flow rates and is a smaller, more robust plasma than the ICP. Peak broadening is thus less of a concern. Disadvantages include the necessity to "vent" solvent prior to the MIP owing to plasma instability in the presence of high concentrations of organics. A valve between the column and transfer line is generally used to vent the solvent.[71] There are two main categories of MIP/MS found in the literature: low pressure[72–74,77–79] and atmospheric pressure.[75–79]

7.3.1.1. Low-Pressure GC/MIP/MS

By using reduced-pressure MIPs, polyatomic interferences at low masses may be alleviated. This is advantageous especially for the analysis of elements with interferences from oxygen and nitrogen. Entrainment of air produces such interferences as NOH^+ and O_2^+ at m/z values of 31 and 32 respectively, which interferes with the determination of sulfur and phosphorus.[15] Chlorine and bromine (with molecular masses of 35 and 79, respectively) suffer from oxygen–hydrogen interferences at these m/z values. Possible isobaric interferences at m/z 35 and 37 include $[^{16}O^{18}OH]^+$, $[^{16}O_2H_3]^+$, $[^{16}O_2H_5]^+$, and $[^{36}ArH]^+$. In addition, bromine has a potential interference from nickel oxide and hydroxide at m/z 79 due to nickel present in the sampling cone. Reduction of these interferences is possible through the use of low-pressure MIPs. The use of high-purity helium also contributes to interference reduction.[15]

7.3.1.1.1. Applications. A low-pressure MIP/MS system was used by Markey and Abramson to convert carbon-containing compounds in a capillary GC effluent to CO and CO_2.[79] Through the use of low operating pressures, no makeup gas was necessary for this study. Complete combustion of analyte introduced was accomplished with only 0.1 mL/min of O_2. By use of a sample of methyl[1-^{14}C]palmitate, the potential of the technique as a monitor of radioactivity was assessed. Markey and co-worker suggest the potential usefulness of the technique in metabolic studies or in survey analyses.

Heppner discussed the development of a low-pressure MIP system for selective elemental detection in combination with GC/MS.[78] The technique was used to convert complex organic molecules into a few simple neutral species. Speciation of these neutrals allowed the determination of elemental composition information for the original molecules. Elemental carbon, nitrogen, oxygen, sulfur, and chlorine were detected, and the elemental ratios of C/O and C/N for various compounds were determined.

A low-pressure helium MIP/MS system was employed as a detector for halogenated GC effluents.[74] Creed et al. optimized the system with respect to power and first-stage pressure. Preliminary detection limits for iodine (0.1 pg), bromine (3.5 pg), and chlorine (24 pg) were reported. Two interface designs were investigated. As discussed above in Section 7.3.1.1, ablation of the sampler resulted in the formation of undesirable background species. Elimination of this background was investigated with the use of an aluminum interface as opposed to the nickel interface originally employed.

As previously stated, low-pressure plasmas reduce the formation of undesirable background species occurring at the major isotopes of phosphorus and sulfur.[73] Both reduced-pressure helium and nitrogen MIPs were evaluated as ion sources by Story et al. The detection limit for phosphorus in triethyl phosphite was found to depend on the amount of torch cooling employed for the helium MIP. The range of detection limits was from 1 to 90 ng. The use of an air-cooled jacket increased torch cooling and improved absolute detection. The use of the nitrogen plasma gave better signal and less interaction of phosphorus with the torch walls in comparison to the helium source. This was thought to be a consequence of the reactivity of phosphorus toward the quartz torch used for helium plasma determinations. Sub-ng limits of detection were obtained by using a nitrogen plasma for both phosphorus and

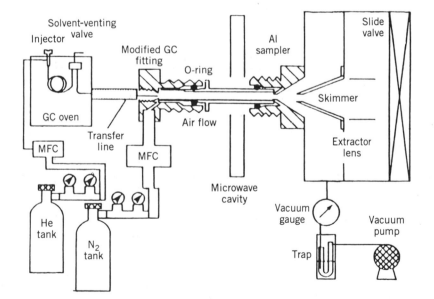

Figure 7.6. Schematic diagram of a low-pressure MIP/MS interface (MFC = mass flow controller). Reprinted from Story et al.,[73] with permission of the Royal Society of Chemistry.

sulfur in pesticides (diazinon and malathion). Figure 7.6 shows a schematic diagram detailing the reduced-pressure interface used in this study.

A recent report by Story and Caruso described the use of GC/MIP/MS at reduced MIP pressures for the determination of phosphorus, sulfur, and the halogens.[72] Three improvements over the previously reported system were evaluated: first, a water-cooled torch was used in this study to improve sensitivity; second, a small amount of hydrogen was added to the plasma gas to serve both as a scavenger of phosphorus before reaction with the quartz torch and to reduce polyatomic ion formation; and third, the interface between the GC transfer line and the plasma torch was improved to minimize dead volume and subsequent band broadening. Detection limits were in the sub-ng range for all elements studied (Br, Cl, I, P, and S). A mixture of seven

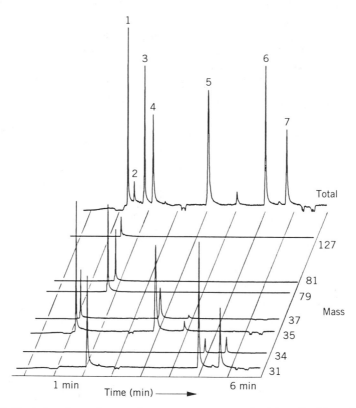

Figure 7.7. Single-ion chromatograms from a seven-component mixture (1, chlorotoluene; 2, iodobenzene; 3, bromononone; 4, triethyl phosphate; 5, chloronaphthalene; 6, diazinon; 7, malathion); 32 ng injection of each component. Reprinted from Story and Caruso,[72] with permission of The Royal Society of Chemistry.

compounds was analyzed simultaneously for the latter elements at their major isotopes (except sulfur): m/z 79 and 81 for bromine; m/z 35 and 37 for chlorine; m/z 127 for iodine; m/z 31 for phosphorus; and, finally, m/z 34 for sulfur. Figure 7.7 shows single-ion chromatograms from the seven-component mixture, as well as the total ion chromatogram.

Both an atmospheric pressure and a low-pressure helium MIP/MS system were utilized by Creed and his colleagues for the detection of halogenated compounds.[77] A capillary gas chromatograph was used in the atmospheric pressure system with picogram sensitivities for both bromine and iodine. Elevated backgrounds at m/z 35 and 37 (the major isotopes of chlorine) led to the design of the low-pressure torch, which successfully reduced these as well as other backgrounds occurring at low masses (by reducing atmospheric entrainment).

7.3.1.2.   Atmospheric Pressure GC/MIP/MS

The simplicity of the instrumental design is an advantage of atmospheric pressure plasmas. As noted earlier, however, entrainment of air in the plasma produces unwanted polyatomic ions that interfere with certain analyses. Atmospheric pressure MIPs with MS detection, therefore, have had limited applications. A few applications are reviewed next.

7.3.1.2.1.   Applications. Capillary gas chromatographic detection for the speciation of organotin compounds was used by Suyani et al. with helium MIP/MS detection.[75] A slightly polar DB-1701 stationary phase was employed for this study, and two different tangential flow torches were evaluated. Tin detection limits for trialkyltin and tetraalkyltin compounds were lower with the use of a tantalum injector torch (0.09–0.35 pg) in comparison to a standard tangential flow torch (1–4 pg). The reduction in detection limits was a result of the introduction of the GC effluent directly from the capillary column into the helium plasma, which confined the effluent in the center of the plasma.

Halogenated hydrocarbons (chlorobenzene, p-chlorotoluene, o-chlorophenol, 2-bromooctane, 1-bromononane, and iodobenzene) were determined by capillary GC/MIP/MS by Mohamad and co-workers.[76] Detection limits (in pg) for these compounds were 9.2, 11.0, 21.0, 1.08, 0.92, and 1.5, respectively. Detection limits for chlorinated compounds were higher owing to elevated backgrounds at both of the major isotopes of chlorine. A 0.4 mm sampling orifice was employed to reduce background, presumably by reducing atmospheric entrainment. A commercial argon ICP/MS system was modified to meet the needs of a helium MIP/MS system.

7.3.2. GC/ICP/MS

Chong and Houk employed a packed GC column for the ICP/MS determinations of organic compounds.[83] Elemental ratios were determined, and the range of detection limits was $0.001-400 \, ng \, s^{-1}$. The units of $ng \, s^{-1}$ were used since a transient chromatographic signal was detected and the authors wished to facilitate the comparison of GC columns of various efficiencies under differing temperatures. Chong and Houk concluded that GC/ICP/MS provided elemental and isotope ratio information that was complementary to the information obtained from GC/MS analyses.

Kim and colleagues constructed a capillary GC/ICP/MS transfer line with only simple modifications being required.[81] Analysis of organolead species in a complex hydrocarbon mixture (NIST SRM 1637, Tetraethyllead Motor Mix in Reference Fuel) was reported. The GC/ICP/MS ion chromatogram taken at m/z 208 of a naphtha sample containing five organolead components is shown in Figure 7.8. Detection limits of 0.7 pg/s were achieved, and the

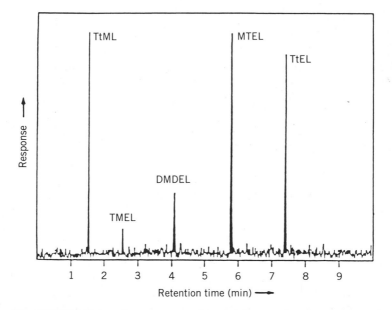

Figure 7.8. GC/ICP/MS ion chromatogram of a naphtha sample containing five alkyllead components:TtEL tetraethyllead; TtML tetramethyllead; TMEL trimethylethyllead; DMDEL dimethyldiethyllead; MTEL methyltriethyllead. Reprinted from Kim et al.,[81] with permission of The Royal Society of Chemistry.

authors suggest the applicability of their method to the analysis of relatively involatile organometallics.

The aforementioned methodology was applied to the analysis of volatile organometallic compounds containing iron, nickel, and tin by Kim et al. using capillary GC/ICP/MS.[82] The researchers analyzed a mixture of organotin compounds in spiked water samples and confirmed the presence of volatile organotin compounds in a sample of harbor sediment. Nickel diethyldithiocarbamate and ferrocene in hexane solutions were also studied. The detection of the latter compound is generally not possible by HPLC/ICP/MS owing to an ArO^+ interference at m/z 56 that corresponds to ^{56}Fe. The former compound is relatively strongly retained, but its elution was facilitated through the use of resistive heating of the stainless steel tube (through which the column passed).

A recent report by Peters and Beauchemin characterized an interface that allows either nebulization or GC as sample introduction systems in ICP/MS.[80] Sensitivity, detection limits, and signal-to-noise ratios (S/N) for aqueous solutions of various elements were investigated in the nebulization mode. A heated interface coupling the spray chamber and the plasma torch resulted in an improvement in performance. Also, detection limits were lower than with the use of a conventional setup. The heated interface was evaluated for the GC determination of two chlorinated compounds, and GC studies of tin compounds were additionally accomplished. The researchers concluded that, on the basis of improved detection limits, the system may permanently remain on the ICP/MS.

7.3.3. Future Directions

Current GC/plasma/MS work in our laboratory has involved the use of a low-pressure ICP/MS system for the determination of halogenated compounds. The difference between a low-pressure ICP and an atmospheric ICP rests in the use of reduced powers and flow rates with the former, but the same radio-frequency (RF) generator and torch box design of the latter may still be used. This is an interesting alternative to the detection schemes given above.

The advantages of GC/plasma detection over HPLC/ICP/MS include nearly 100% transport efficiency, fewer isobaric interferences, reduced wear of sampler and skimmer cones, and good plasma stability.[82] Chromatographic resolution by capillary GC/ICP/MS is better than with packed columns.[82] Disadvantages include the inability to analyze as wide a variety of species as may be determined by HPLC/ICP/MS. Nevertheless, GC/plasma MS may have important applicability, especially in the analysis of many environmental samples containing pesticides.

7.4. SUPERCRITICAL FLUID CHROMATOGRAPHY

A supercritical fluid is a gas-like, compressible fluid heated and pressurized above its critical temperature and pressure. Supercritical fluid chromatography (SFC) was introduced as a technique with great potential more than 30 years ago.[87] SFC is thought to be a combination of the best features HPLC and GC have to offer, e.g., the solvating power of liquids and the speed of analysis of GC. As a result, interest in SFC has increased in recent years.

Advantages of SFC over GC and HPLC include the following: thermally labile, nonvolatile, and high molecular weight compounds may be analyzed by SFC but not by GC; diffusion constants of supercritical fluids are greater than those of liquids, resulting in a lower resistance to mass transfer in the mobile phase, thereby shortening analysis times and improving resolution of complex mixtures as compared to HPLC[88]; additionally, the "general elution problem" may be overcome by SFC. The general elution problem states that compounds with large k' values [$k' =$ (retention time of the analyte of interest $-$ the retention time of the solvent) $\div$ the retention time of the solvent] give broad, highly retained peaks. Conversely, small k' values generally result in poor separation.[89] With SFC, retention is a complex function of many variables, including temperature, pressure, density, sample concentration, composition of the mobile phase, and composition of the stationary phase. Temperature programming is used in GC to alleviate the general elution problem, and gradient elution is used for the same purpose in HPLC to optimize separation conditions. In SFC, however, it is possible to vary temperature, mobile phase strength, and density or pressure to change solvating properties, and thus k'. Subsequently, SFC is more versatile than HPLC and GC in this regard.[89]

Modifiers such as acetonitrile, chloroform, methanol, and water are required when somewhat polar compounds, especially those having limited solubility in nonpolar mobile phases, are to be separated in supercritical carbon dioxide. Modifiers are used to alter the solvent strength of the supercritical mobile phase. The partition coefficients of the analyte of interest between the solvent and the stationary phase are also changed by the addition of polar modifiers.[90]

One of the most popular SFC detectors has been the flame-ionization detector (FID). The FID has also been extensively studied as a GC detector; it is simple and easy to operate, inexpensive, and responds sensitively to many types of combustible compounds. FID detection, however, lacks the ability to distinguish between coeluting compounds. Accordingly, SFC has been coupled to both MIP/MS and ICP/MS detectors, thereby producing high selectivity. ICP has the advantage over MIP in that polar modifiers may be used to change mobile phase polarity without apparent effect on the

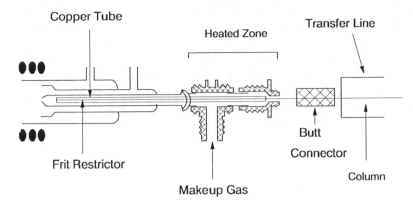

Figure 7.9. SFC/ICP/MS interface. Reprinted from Vela and Caruso,[91] with permission of Elsevier Science Publishers B.V.

plasma.[89,91] As with any new coupling scheme, however, various problems have arisen.

Carbon dioxide is the most frequently used supercritical fluid owing to its low critical temperature and pressure. It is inert, nontoxic, reasonably priced, and available in a fairly pure state. Nitrous oxide and ammonia are possible alternatives to CO_2, but the former is highly reactive and the latter is corrosive.[92] One of the disadvantages of SFC/ICP/MS is that the introduction of CO_2 (or other mobile phases) reduces sensitivity to a certain extent by quenching the plasma.

Table 7.4. Elements Investigated by SFC/Plasma MS[a]

Element	Plasma Type	Absolute Detection Limit (pg)	References
Br	MIP	0.75	94
Cl	MIP	15	94
Cr	ICP	0.9–3	97
Pb	ICP	0.5–100	96
Hg	ICP	3–10	96
Sn	ICP	0.03–0.8	91, 93, 95

[a]No attempt has been made to normalize the detection limits for the compound types. All detection limits are based on 3 SD calculations.

Interfacing the MIP or ICP torch to the restrictor of the SFC has been difficult. Figure 7.9 shows an SFC/ICP/MS interface. Control of fluid flow in SFC is provided by the restrictor. The purpose of the restrictor is to maintain the supercritical conditions through the column and also to maintain a constant mobile phase linear velocity. Heating the restrictor has been found to prevent the column eluent from freezing at the restrictor tip owing to the rapid decompression of the mobile phase, resulting in large changes in baseline signal.[89] Elements investigated by SFC/ICP/MS are given in Table 7.4.

7.4.1. Applications

Shen and co-workers evaluated ICP/MS as an elemental detector for SFC.[93] Organotin compounds were separated with detection levels in the sub-pg range [0.034 pg for tetrabutyltin (TtBT) and 0.047 pg for tetraphenyltin (TtPT)]. These compounds were chosen to explore the effects of varying SFC parameters. Time-resolved aquisition (TRA) software was used to show the potential for multimass monitoring for mass-selective chromatograms (for a single injection). Figure 7.10 shows a TRA acquisition for a scan of 100 pg of TBT and TPT.

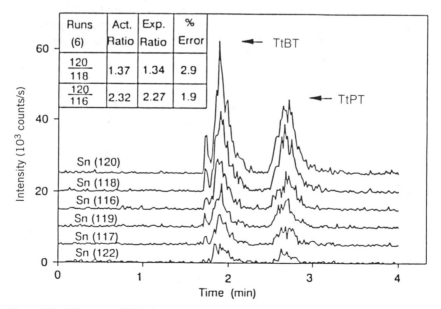

Figure 7.10. TRA scan of SFC chromatograms of various Sn isotopes. Reprinted with permission from Shen et al.[93] Copyright 1991, American Chemical Society.

SFC/helium MIP/MS was evaluated by Olson et al. for the determination of halogenated compounds.[94] Two test compounds, 1-chloronaphthalene and 1-bromo-2-methylnaphthalene, were employed for this study. Absolute detection limits were determined to be 15 pg for Cl and 0.75 pg for Br. The interface temperature of the SFC/MIP frit restrictor must be maintained at 100 °C. A transfer line was used to connect the SFC to the tangential flow torch. TRA software was employed to simultaneously detect chlorine and bromine

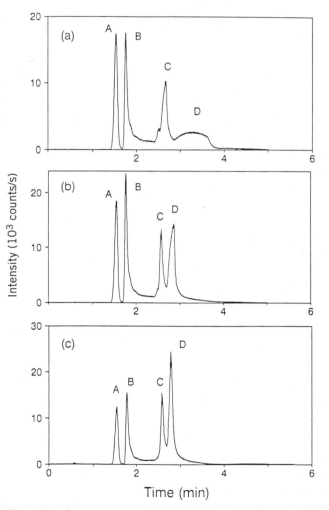

Figure 7.11. Effect of interface temperature (a) 215, (b) 300, and (c) 350 °C on SFC separations of a mixture containing (A) TtBT, (B) TBT–Cl, (C) TPT–Cl, and (D) TtPT. Reprinted from Vela and Caruso,[95] with permission of The Royal Society of Chemistry.

in the compounds of interest. Additionally, preliminary work on a standard mixture of 15 chlorinated pesticide compounds was done.

Organotin compounds are not readily amenable to separation by GC or HPLC. SFC has been utilized by Vela et al. for the determination of organotin compounds of differing polarities (tri- and tetraorganotin compounds) separated on a polarizable stationary phase using ICP/MS detection.[95] The effect of carbon dioxide introduction to an argon plasma was evaluated, and the effect of the restrictor temperature determined. Figure 7.11 shows the effect of interface temperature on SFC chromatograms of TtBT, tributyltin chloride (TBT–Cl), triphenyltin chloride (TPT–Cl), and TtPT. Also, the effect of different parameters on the separation of the organotin compounds was discussed. Tin detection limits were found to be 0.26 pg for TtBT, 0.80 pg for TBT–Cl, 0.57 pg for TPT–Cl, and 0.20 pg for TtPT.

Organolead and organomercury compounds were investigated by Carey and co-workers to evaluate the feasibility of SFC/ICP/MS as a multielement detection scheme.[96] TRA acquisitions were compared with data from single-ion monitoring. For diethylmercury (DIEM), tetrabutyllead (TtBL), and tributyllead (TBL) acetate, detection limits were lower when the single-ion monitoring mode was used. Low to sub-picogram detection limits were obtained by using both modes for all three compounds, and the NIST SRM 2715 Lead in Reference Fuel, was used to verify results from single-ion monitoring studies.

A comparison of FID and ICP/MS detectors for the determination of organometallic compounds (TtBT, TtPT, TBT-Cl, TPT-Cl) separated by capillary SFC was made by Vela and Caruso.[91] The resolution achieved with SFC/FID detection was not always seen with SFC/ICP/MS detection; peak broadening was attributed to fluctuations in the transfer line temperature. The use of a longer column gave baseline resolution for the organotin compounds of interest for both detection techniques. An order of magnitude improvement in detection limits was achieved by using SFC/ICP/MS (sub-pg level for all species studied). The relative standard deviations (RSDs) for five tin injections were better when SFC/ICP/MS detection was used.

7.4.2. Future Developments

As is obvious by the small number of published reports discussing the coupling of SFC to plasma mass spectrometry, SFC/plasma MS is in an early developmental stage. All of the SFC/ICP/MS papers deal with the separation of organometallic compounds. Since a major area of plasma spectroscopic analysis involves inorganic metal ions, the inability thus far to effectively determine these or highly polar analytes represent prominent weaknesses of the technique.

Possible solutions include the use of polar or reversed-micellar mobile phases or the use of polar modifiers in nonpolar mobile phases. These three areas, along with interfacial studies, require further development. Nevertheless, SFC does appear promising in the world of chromatographic trace elemental speciation. In our laboratory, recent studies have been performed on the separation of organochromium compounds by SFC/ICP/MS.[97]

7.5. CONCLUSIONS

It is clear from the foregoing discussion that the analysis and speciation of trace elements in "real world" samples is an important area of interest to many researchers. There are many pitfalls to chromatographic analysis, and future studies in the coupling of HPLC, GC, and SFC to ICP/MS will surely address these. Despite the problems that may arise, chromatographic sample analysis with plasma MS detection is proving to be a viable marriage in the world of analytical chemistry.

7.6. SUMMARY

Plasma source mass spectrometry offers selective detection with excellent sensitivity. When coupled with chromatographic techniques such as HPLC (high-performance liquid chromatography), GC (gas chromatography), and SFC (supercritical fluid chromatography), it provides a method for elemental speciation. A review of investigations coupling chromatography with both ICP/MS (inductively coupled plasma–mass spectrometry) and MIP/MS (microwave-induced plasma–mass spectrometry) detectors is presented. For HPLC/ICP/MS detection, most reported detection limits are in the sub-ng to pg range. Gaseous sample introduction techniques generally provide better detection limits, with levels ranging from pg to sub-pg.

Acknowledgments

We are grateful to the National Institute of Environmental Health Sciences for partial support of this work through grant numbers ES03321 and ES04908. We acknowledge the NIH-BRS Shared Instruments Grants program for providing the VG PlasmaQuad through grant number S10RR02714 and the U.S. Environmental Protection Agency for partial support of this work through grant number CR-818301.

REFERENCES

1. C. Veillon, *Anal. Chem.* **58**, 851A–866A (1986).
2. D. Heitkemper, J. Creed, J. A. Caruso, and F. L. Fricke, *J. Anal. At. Spectrom.* **4**, 279–284 (1989).
3. B. S. Sheppard and J. A. Caruso, *Analyst (London)* **117**, 971–975 (1992).
4. R. A. Goyer, in *Casarett and Doull's Toxicology: The Basic Science of Poisons* (C. D. Klaassen, M. D. Amdur, and J. Doull, eds.), 3rd ed., Chapter 19. Macmillan, New York, 1986.
5. H. M. Crews, J. R. Dean, L. Ebdon, and R. C. Massey, *Analyst (London)* **114**, 895–899.
6. D. T. Gjerde and H. C. Mehra, in *Trace Metal Analysis and Speciation* (I. S. Krull, ed.), J. Chromatogr. Libr., Vol. 47, Chapter 8. Elsevier, Amsterdam, 1991.
7. H. Salem and S. A. Katz, eds., *Sci. Total Environ.* **86**, 1–206 (1989).
8. *Toxicological Profile for Chromium.* Oak Ridge Nat. Lab., ed. Oak Ridge, TN, 1989. Prepared by Syracuse Research Corporation for the Agency for Toxic Substances and Disease Registry, U.S. Public Health Service.
9. R. C. Elder, Z. Zhao, Y. Zhang, J. G. Dorsey, E. V. Hess, and K. Tepperman, *J. Rheumatol.* **20**, 268–272 (1993).
10. Z. Zhao, W. B. Jones, K. Tepperman, J. G. Dorsey, and R. C. Elder, *J. Pharm. Biomed. Anal.* **10**, 279–287 (1992).
11. A. Al-Rashdan, D. Heitkemper, and J. A. Caruso, *J. Chromatogr. Sci.* **29**, 98–102 (1991).
12. A. Al-Rashdan, N. P. Vela, and J. A. Caruso, *J. Anal. At. Spectrom.* **7**, 551–555 (1992).
13. U. T. Kumar, J. G. Dorsey, J. A. Caruso, and E. H. Evans, *J. Chromatogr. A*, **654**, 261–268 (1993).
14. A. Montaser and D. W. Golightly, eds., *Inductively Coupled Plasmas in Analytical Spectrometry*, 2nd ed., VCH Publishers, New York, 1992.
15. J. M. Carey, F. A. Byrdy, and J. A. Caruso, *J. Chromatogr. Sci.* **31**, 330–344 (1993).
16. E. H. Evans and J. J. Giglio, *J. Anal. At. Spectrom.* **8**, 1–18 (1993).
17. D. T. Heitkemper and J. A. Caruso, in *Trace Metal Analysis and Speciation* (I. S. Krull, ed.), J. Chromatogr. Libr., Vol. 47, Chapter 3. Elsevier, Amsterdam, 1991.
18. L. K. Olson, D. T. Heitkemper, and J. A. Caruso, in *Element-Specific Chromatographic Detection by Atomic Emission Spectroscopy* (P. C. Uden, ed.), ACS Symp. Ser. Vol. 479, Chapter 17. American Chemical Society, Washington, D.C., 1992.
19. U. Kumar, J. G. Dorsey, J. A. Caruso, and E. H. Evans, *J. Chromatogr. Sci.* **32**, 282–285 (1994).
20. S. Branch, L. Ebdon, and P. O'Neill, *J. Anal. At. Spectrom.* **9**, 33–37 (1994).
21. U. T. Kumar, N. P. Vela, J. G. Dorsey, and J. A. Caruso, *J. Chromatogr. A*, **655**, 340–345 (1993).
22. W. Pretorius, M. Foulkes, L. Ebdon, and S. Rowland, *J. High Res. Chromatogr.* **16**, 157–160 (1993).
23. X. Dauchy, R. Cottier, A. Batel et al. *J. Chromatogr. Sci.* **31**, 416–421 (1993).
24. S. C. K. Shum and R. S. Houk, *Anal. Chem.* **65**, 2972–2976 (1993).

25. Z. Zhao, K. Tepperman, J. G. Dorsey, and R. C. Elder, *J. Chromatogr. Biomed. Appl.* **615**, 83–89 (1993).

26. K. Takatera and T. Watanabe, *Anal. Chem.* **65**, 759–762 (1993).

27. K. Takatera and T. Watanabe, *Anal. Sci.* **9**, 19–23 (1993).

28. E. J. Arar, S. E. Long, T. D. Martin, and S. Gold, *Environ. Sci. Technol.* **26**, 1944–1950 (1992).

29. D. S. Braverman, *J. Anal. At. Spectrom.* **7**, 43–46 (1992).

30. J. S. Edmonds, Y. Shibata, K. A. Francesconi, J. Yoshinaga, and M. Morita, *Sci. Total Environ.* **122**, 321–335 (1992).

31. R. C. Elder, W. B. Jones, and K. Tepperman, in *Element-Specific Chromatographic Detection by Atomic Emission Spectroscopy* (P. C. Uden, ed.), ACS Symp. Ser. Vol. 479, Chapter 18. American Chemical Society, Washington, D.C., 1992.

32. S. Hill, *Anal. Proc. (London)* **29**, 399–401 (1992).

33. S. Honoré Hansen, E. H. Larsen, G. Pritzl, and C. Claus, *J. Anal. At. Spectrom.* **7**, 629–634 (1992).

34. L. M. W. Owen, H. M. Crews, R. C. Hutton, and A. Walsh, *Analyst (London)* **117**, 649–655 (1992).

35. S. C. K. Shum, H. Pang, and R. S. Hauk, *Anal. Chem.* **64**, 2444–2450 (1992).

36. S. C. K. Shum, R. Neddersen, and R. S. Houk, *Analyst (London)* **117**, 577–582 (1992).

37. K. Takatera and T. Watanabe, *Anal. Sci.* **8**, 469–474 (1992).

38. L. M. W. Owen, H. M. Crews, and R. C. Massey, *Chem. Speciation Bioavailability* **4**, 89–96 (1992).

39. B. Gercken and R. M. Barnes, *Anal. Chem.* **63**, 283–287 (1991).

40. R. S. Houk, S.-J. Jiang, in *Trace Metal Analysis and Speciation* (I. S. Krull, ed.), J. Chromatogr. Libr., Vol. 47, Chapter 5. Elsevier, Amsterdam, 1991.

41. K. Kawabata, Y. Kishi, O. Kawaguchi, Y. Watanabe, and Y. Inoue, *Anal. Chem.* **63**, 2137–2140 (1991).

42. J. M. Riviello, A. Siriraks, R. M. Manabe, R. Roehl, and M. Alforque, *LC-GC* **9**, 704–712 (1991).

43. K. Takatera and T. Watanaber, *Anal. Sci.* **7**, 695–698 (1991).

44. D. W. Boomer, M. J. Powell, and J. Hipfner, *Talanta* **37**, 127–134 (1990).

45. D. Heitkemper, J. Creed, and J. A. Caruso, *J. Chromatogr. Sci.* **28**, 175–181 (1990).

46. H. Klinkenberg, S. van der Wal, J. Frusch, L. Terwint, and T. Beeren, *At. Spectrosc.* **11**, 198–201 (1990).

47. A. Z. Mason, S. D. Storms, and K. D. Jenkins, *Anal. Biochem.* **186**, 187–201 (1990).

48. R. Roehl and M. M. Alforque, *At. Spectrosc.* **11**, 210–215 (1990).

49. B. S. Sheppard, W. L. Shen, J. A. Caruso, D. T. Heitkemper, and F. L. Fricke, *J. Anal. At. Spectrom.* **5**, 431–435 (1990).

50. D. Beauchemin, K. W. M. Sium, J. W. McLaren, and S. S. Berman, *J. Anal. At. Spectrom.* **4**, 285–289 (1989).

51. S. Branch, L. Ebdon, S. Hill, and P. O'Neill, *Anal. Proc. (London)* **26**, 401–403 (1989).

52. M. A. Brown, I. S. Kim, R. Roehl, F. I. Sasinos, and R. D. Stephens, *Chemosphere* **19**, 1921–1927 (1989).

53. D. S. Bushee, J. R. Moody, and J. C. May, *J. Anal. At. Spectrom.* **4**, 773–775 (1989).

54. A. Dyer, *Analyst (London)* **114**, 265–267 (1989).
55. A. Z. Mason, *Chromatogram* **10**, 7–10 (1989).
56. S. G. Matz, R. C. Elder, and K. Tepperman, *J. Anal. At. Spectrom.* **4**, 767–771 (1989).
57. Y. Shibata and M. Morita, *Anal. Sci.* **5**, 107–109 (1989).
58. H. Suyani, J. Creed, T. Davidson, and J. A. Caruso, *J. Chromatogr. Sci.* **27**, 139–143 (1989).
59. H. Suyani, D. Heitkemper, J. Creed, and J. A. Caruso, *Appl. Spectrosc.* **43**, 962–967 (1989).
60. D. Beauchemin, M. E. Bednas, S. S. Berman, J. W. McLaren, K. M. W. Siu, and R. E. Sturgeon, *Anal. Chem.* **60**, 2209–2212 (1988).
61. D. S. Bushee, *Analyst (London)* **113**, 1167–1170 (1988).
62. H. M. Crews, R. Massey, D. J. McWeeny, and J. R. Dean, *J. Res. Natl. Bur. Stand.* **93**, 349–354 (1988).
63. S. J. Jiang and R. S. Houk, *Spectrochim. Acta* **43B**, 405–411 (1988).
64. S. I. Suzuki, H. Tsuchihashi, K. Nakajima, A. Matsushita, and T. Nagao, *J. Chromatogr.* **437**, 322–327 (1988).
65. J. R. Dean, S. Munro, L. Ebdon, H. M. Crews, and R. C. Massey, *J. Anal. At. Spectrom.* **2**, 607–610 (1987).
66. J. J. Thompson, R. S. Houk, *Anal. Chem.* **58**, 2541–2548 (1986).
67. F. A. Byrdy and J. A. Caruso, *Environ. Health Perspect.* (in press) (1995).
68. L. R. Synder and J. J. Kirkland, *Introduction to Modern Liquid Chromatography*, 2nd ed. Wiley, New York, 1980.
69. J. M. Miller, *Chromatography: Concepts and Contrasts*. Wiley, New York, 1988.
70. Analytical Methods Documentation, Ontario Ministry of the Environment, Trace Metal Analyses of Acid Precipitation, Document 00678 (1987).
71. N. P. Vela, L. K. Olson, and J. A. Caruso, *Anal. Chem.* **65**, 585A–597A (1993).
72. W. C. Story, J. A. Caruso, *J. Anal. At. Spectrom.* **8**, 571–575 (1993).
73. W. C. Story, L. K. Olson, W. L. Shen, J. T. Creed, and J. A. Caruso, *J. Anal. At. Spectrom.* **5**, 467–470 (1990).
74. J. T. Creed, T. M. Davidson, W. L. Shen, and J. A. Caruso, *J. Anal. At. Spectrom.* **5**, 109–113 (1990).
75. H. Suyani, J. Creed, J. A. Caruso, and R. D. Satzger, *J. Anal. At. Spectrom.* **4**, 477–782 (1989).
76. A. H. Mohamad, J. T. Creed, T. M. Davidson, and J. A. Caruso, *Appl. Spectrosc.* **43**, 1127–1131 (1989).
77. J. T. Creed, A. H. Mohammed, T. M. Davidson, G. Ataman, and J. A. Caruso, *J. Anal. At. Spectrom.* **3**, 923–926 (1988).
78. R. A. Heppner, *Anal. Chem.* **55**, 2170–2174 (1983).
79. S. P. Markey and F. P. Abramson, *Anal. Chem.* **54**, 2375–2376 (1982).
80. G. R. Peters and D. Beauchemin, *Anal. Chem.* **65**, 97–103 (1993).
81. A. W. Kim, M. E. Foulkes, L. Ebdon, S. J. Hill, R. L. Patience, A. G. Barwise, and S. J. Rowland, *J. Anal. At. Spectrom.* **7**, 1147–1149 (1992).
82. A. Kim, S. Hill, L. Ebdon, and S. Rowland, *J. High Resolut. Chromatogr.* **15**, 665–668 (1992).
83. N. S. Chong and R. S. Houk, *Appl. Spectrosc.* **41**, 66–74 (1987).

84. A. J. McCormack, S. C. Tong, and W. D. Cooke, *Anal. Chem.* **37**, 1470–1476 (1965).

85. A. H. Mohamad and J. A. Caruso, *Adv. Chromatogr.* **26**, 191 (1987).

86. D. J. Douglas and J. B. French, *Anal. Chem.* **53**, 37–41 (1981).

87. E. Klepser, A. H. Corwin, and D. A. Turner, *J. Org. Chem.* **27**, 700–701 (1962).

88. R. D. Smith, B. W. Wright, and C. R. Yonker, *Anal. Chem.* **60**, 1323A–1336A (1988).

89. J. M. Carey and J. A. Caruso, *Trends Anal. Chem.* **11**, 287–293 (1992).

90. N. P. Vela and J. A. Caruso, *J. Anal. At. Spectrom.* **8**, 787–794 (1993).

91. N. P. Vela and J. A. Caruso, *J. Chromatogr.* **641**, 337–345 (1993).

92. R. E. Majors, *LC-GC* **9**, 78–86 (1991).

93. W. L. Shen, N. P. Vela, B. S. Sheppard, and J. A. Caruso, *Anal. Chem.* **63**, 1491–1496 (1991).

94. L. K. Olson and J. A. Caruso, *J. Anal. At. Spectrom.* **7**, 993–998 (1992).

95. N. P. Vela and J. A. Caruso, *J. Anal. At. Spectrom.* **7**, 971–977 (1992).

96. J. M. Carey, N. P. Vela, and J. A. Caruso, *J. Anal. At. Spectrom.* **7**, 1173–1181 (1992).

97. J. M. Carey and J. A. Caruso, *J. Chromatogr.* **622**, 329–340 (1994).

CHAPTER

8

PEROXYOXALATE CHEMILUMINESCENCE: MECHANISM AND ANALYTICAL DETECTION

ANDREW G. HADD and JOHN W. BIRKS

*Department of Chemistry
and Cooperative Institute for Research in Environmental Sciences (CIRES),
University of Colorado at Boulder,
Boulder, Colorado, 80309-0216*

In recent years, chemiluminescence (CL) has become a powerful analytical tool for sensitive and selective detection of chemical species. The chemiluminescence technique, which is based on light generated in a chemical reaction, demonstrates significant advantages over other detection methods such as fluorescence. The major limitation of fluorescence detection, the background noise associated with scattered light from the excitation source, is absent in CL detection, often allowing for more sensitive detection. Another advantage of chemiluminescence over fluorescence is the ability to detect analytes in complex matrices with greater selectivity. CL techniques can also be used for the detection of compounds that have no chromophores and therefore are not detected by either UV absorbance or fluorescence. Instruments based on chemiluminescence have the additional advantage of being relatively simple in construction and inexpensive.

This chapter focuses on peroxyoxalate chemiluminescence (PO–CL), the most efficient nonenzymatic CL reaction known. PO–CL forms the basis for Cyalume® "light sticks," which are used for emergency lighting, for light sources in medical examinations, and for novelty items such as the glow necklaces sold at amusement parks. Because of its high efficiency, the PO–CL reaction has been developed as a sensitive technique for a wide range of analytical problems.[1] Whereas most chemiluminescence reactions involve emission from a reaction intermediate derived from one of the reagents, the peroxyoxalate reaction transfers energy to a variety of fluorescent molecules, which in turn emit light identical to their fluorescence. PO–CL is currently

Selective Detectors, edited by Robert E. Sievers. Chemical Analysis Series, Vol. 131.
ISBN 0-471-01343-9 © 1995 John Wiley & Sons, Inc.

the most sensitive and versatile chemiluminescence detection method for liquid chromatography (LC). In a number of cases, a 10- to 100-fold increase in sensitivity over conventional fluorescence detection has been demonstrated.[2-5] Another feature of PO–CL detection is that any of the reagents, i.e., fluorophore, H_2O_2, base catalyst, or oxalate, may serve as the analyte.

In this chapter we begin by attempting to provide the reader with a fundamental understanding of the peroxyoxalate reaction mechanism, as it is presently understood. The reaction involves the formation of a high energy intermediate (or intermediates) capable of transferring energy to fluorescent molecules. Further research of the PO–CL mechanism, which may lead to improved detection limits and novel applications, is still needed. The practical aspects of designing a CL method for detection in flowing systems, such as high-performance liquid chromatography (HPLC), are also discussed. We describe the reaction variables affecting CL intensity, mixing of the CL reagents, and common instrument designs. Finally, examples of sensitive and selective detection techniques based on PO–CL are discussed. These examples demonstrate the versatility of using the PO–CL reaction for analytical applications.

8.1. REACTION MECHANISM

Chandross first reported the chemiluminescence from oxalyl chloride and hydrogen peroxide in the presence of fluorescent molecules in 1963.[6] As investigators sought long-lived light sources for commercial and military uses, related chemiluminescence studies were carried out using phenyl oxalate esters.[7,8] The reaction was found to be highly efficient and capable of exciting a wide variety of acceptor fluorescent molecules, according to the following general reaction scheme:

$$\text{oxalate ester} + \text{hydrogen peroxide} \rightarrow \text{intermediate (I)} + \text{products}$$

$$\text{I} + \text{fluorophore (F)} \rightarrow \text{F*} + \text{products}$$

$$\text{F*} \rightarrow \text{F} + h\nu$$

The reaction of an oxalate ester with hydrogen peroxide produces at least one, but probably two or more, highly energetic intermediates capable of generating excited singlet states of fluorescent molecules. Radiative decay of the singlet-excited fluorescent molecule results in emission of light having a spectrum identical to that of its fluorescence. Initial studies found phenyloxalate esters with electronegative substituents to have the greatest chemiluminescent efficiencies,[9] with chemiluminescence quantum yields, Φ_{CL}, of up to

23%. The PO–CL reaction is applicable to a wide range of fluorescent molecules, with an energy transfer ability of up to $430\,kJ/mol$.[10] The versatility and high-energy characteristics of PO–CL have led to many studies concerning the reaction mechanism and nature of the intermediates involved.

Since the discovery of PO–CL in the early 1960s, the identity of the intermediate(s) responsible for exciting the fluorophore has been debated.[11-13] In their initial work, Rauhut and co-workers proposed 1,2-dioxetanedione (**I**) as the reactive intermediate capable of producing the excited singlet state of a fluorescent molecule.

I

The proposed mechanism, given in Scheme I, is initiated by nucleophilic attack of H_2O_2 on one of the carbonyls of the oxalate ester to substitute —OOH for —OAr. It is presumed that in a cyclization reaction the —OOH displaces the remaining —OAr group to form 1,2-dioxetanedione (**I**). Consis-

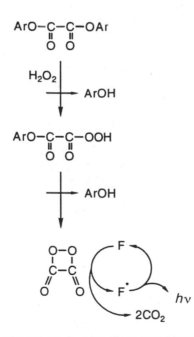

Scheme I. Mechanism for PO–CL as originally proposed by Rauhut and co-workers.[8,9]

tent with this mechanism, the reaction rate increases when —OAr are good leaving groups, i.e., for halogen- and nitro-substituted aromatic rings. Based on this principle, syntheses have been achieved of a large number of oxalate esters that greatly enhance the reaction rate. Two of the more common oxalate esters used for analytical applications are bis(2,4,6-trichlorophenyl)-oxalate (TCPO) and bis(2,4-dinitrophenyl)oxalate (DNPO). The nucleophilic nature of the reaction is also consistent with the observation that the reaction is catalyzed by a wide variety of bases.

The intermediacy of **I** was long favored owing to the analogy with chemiluminescent reactions observed with structurally related dioxetanes (**II**) and dioxetanones (**III**).

II   **III**

These compounds generate singlet excited states in fluorescent molecules by a well-defined chemically intiated electron exchange luminescence (CIEEL) mechanism.[14] The CIEEL mechanism postulated by Schuster involves transfer of an electron from the fluorophore to the reactive intermediate to form a charge transfer complex. The annihilation of oppositely charged radicals in the complex, associated with return of the electron to the fluorophore, is sufficiently energetic to leave the fluorophore in an excited singlet state.

On the basis of Schuster's work, McCapra modified the peroxyoxalate mechanism to include the CIEEL process.[15] Their proposal involves the formation of a short-lived charge transfer complex between 1,2-dioxetanedione and a fluorophore, as shown in Scheme II. The radical anion of 1,2-dioxetanedione is presumed to decompose to form CO_2 and $CO_2^{\cdot-}$. Within the same solvent cage, the electron from $CO_2^{\cdot-}$ is transferred back to the fluorophore radical cation, thereby forming an excited singlet state of the fluorescent molecule. As in fluorescence, internal conversion will rapidly relax the molecule to its first excited singlet state (Kasha's rule). Radiative decay of the excited state produces the observed light emission. The modification of the Rauhut mechanism to include a charge transfer complex (Scheme II) is known as the Rauhut–McCapra mechanism.

In support of the CIEEL mechanism, Sigvardson and Birks[16] and Imai et al.[17] noted a general trend of increasing CL efficiency with decreasing oxidation potential of the fluorophore. Also, Sigvardson and Birks[16] reported a greatly enhanced sensitivity of PO–CL toward amino-substituted polycyclic hydrocarbons (amino-PAHs) as compared to the corresponding unsubstituted

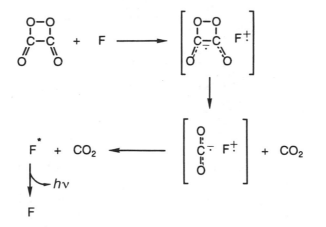

Scheme II. CIEEL mechanism for energy transfer between 1,2-dioxetanedione and the fluorophore as proposed by McCapra.[15]

PAHs. For example, the detection limits for fluoranthene and 1-aminofluoranthene were 15,000 and 0.09 pg, respectively, a difference of more than 5 orders of magnitude. According to the CIEEL mechanism, the charge transfer complex forms more efficiently with compounds having lower oxidation potential, such as amino-PAHs.

Although the CIEEL features of the mechanism have been demonstrated, the presence of dioxetanedione has never been validated despite numerous efforts,[18-20] and it is generally agreed that past sudies reporting the existence of dioxetanedione are flawed in one way or another. For many years the Rauhut–McCapra mechanism remained the working hypothesis despite a lack of independent evidence for dioxetanedione. Recent work has lead to new proposals for the the reaction mechanism and alternative structures of the high-energy intermediate(s) involved.

8.1.1. Kinetics Studies Indicating Multiple Reaction Pathways

In 1984, Catherall et al. carried out the first detailed kinetics study of the PO–CL reaction.[11,21] They studied the reaction of bis(pentachlorophenyl)-oxalate (PCPO) with hydrogen peroxide, using sodium salicylate as a catalyst and 9,10-diphenylanthracene (DPA) as the fluorophore. The following mechanistic pathway summarizes their work.

$$PCPO + H_2O_2 \rightarrow I$$

$$PCPO + H_2O_2 \rightarrow \text{nonchemiluminescent products}$$

$$I \rightarrow \text{nonchemiluminescent products}$$

$$I + F \rightarrow I \cdot F \text{ (charge transfer complex)}$$

$$I \cdot F \rightarrow F^* + \text{products}$$

$$I \cdot F \rightarrow F + \text{products}$$

$$F^* \rightarrow F + h\nu$$

$$F^* \rightarrow F \text{ (radiationless decay)}$$

The rate-determining step of the mechanism is a reaction between the oxalate ester and H_2O_2 and does not involve the fluorophore. An important feature of their work is accounting for side reactions that lead to non-light-producing products. The mechanism explains how small variations in reagent and catalyst concentrations can affect the reaction pathway and thereby the chemiluminescence yield.

Catherall and co-workers also noted a strong dependence of the chemiluminescence efficiency on the structure of the aryl group. In order to account for this relationship, they proposed the cyclization of peroxyoxalate (**IV**) to form the high-energy intermediate **V**.

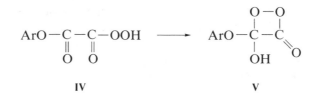

It was proposed that the electron-withdrawing characteristics of the substituted phenol group, ArO—, stabilizes the formation of a charge transfer complex with the fluorophore.

Further evidence for either **IV** or **V** (i.e., an intermediate containing an aryl group) was provided by additional kinetics studies[22] and by fluorine-19 nuclear magnetic resonance (^{19}F-NMR) of the products of the reaction of bis(2,4-difluorophenyl)oxalate with H_2O_2.[23] It should be noted, however, that the existence of either of these intermediates does not rule out 1,2-dioxetanedione (**I**) as an energy transfer intermediate, since **I** could be formed by elimination of the phenol, ArOH, from either **IV** or **V** and could have a lifetime sufficiently short that the chemiluminescence intensity would follow the concentration profiles of its precursors.

In an effort to further explore the reaction mechanism, Alvarez et al. examined the peroxyoxalate reaction under a variety of reaction conditions.[12] In Figure 8.1, for example, the time course for the CL intensity is shown for

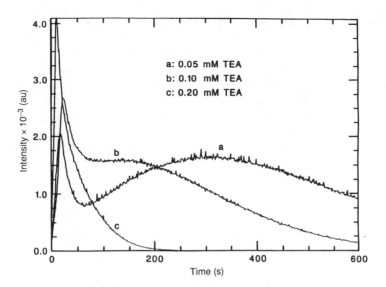

Figure 8.1. Chemiluminescence intensity as a function of time for the reaction of 0.7 mM TCPO with 2 mM H_2O_2 in ethyl acetate and for the concentrations of triethylamine (TEA) given in the figure (a–c). Reprinted from Alvarez et al.,[12] with the permission of the American Chemical Society.

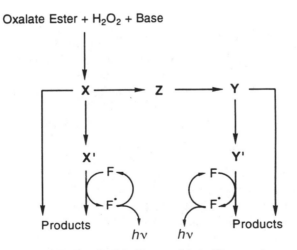

Scheme III. Minimal mechanism for PO-CL according to Givens and co-workers.[12,25]

different concentrations of triethylamine catalyst. The biphasic peak shape at low catalyst concentration reflects multiple reaction pathways. Givens and Schowen reviewed the mathematical aspects in greater detail[24]; their results suggest that more than one intermediate is necessary to account for the biphasic time-intensity profile, as shown in Scheme III. This mechanism, which they describe as the minimal mechanism for PO–CL, requires two intermediates X and Y with a third intermediate, Z, that partitions between them. They also propose that X and Y are not capable of transferring energy to a fluorophore but form energy transfer intermediates X′ and Y′ that complete the scheme. The requirement of formation of X′ and Y′ is based on the zero-order dependence of the reaction rate on fluorophore concentration. As can be seen in Figure 8.1, the timing, duration, and amplitude of the chemiluminescence time profile can be significantly affected by small variations of reagent concentrations. This important consideration for analytical detection is discussed below.

8.1.2.   Mechanistic Information Obtained
from Measurements of Background Emission

A study by Mann and Grayeski of background emission in the absence of a fluorophore for different oxalate esters reacted with hydrogen peroxide provides another confirmation of multiple intermediates.[25] In their experiments, emission spectra were obtained for the reactions of four different oxalate esters with hydrogen peroxide in the presence of a catalyst, but with no fluorophore present. They recorded a weak background emission with a common feature near 450 nm and a longer wavelength emission with features dependent on the oxalate ester used. The longer wavelength emission increased and decreased at a rate faster than the emission at 450 nm, indicative of two light-emitting species. They attributed the 450 nm emission to the phosphorescence of carbon dioxide produced from the decomposition of dioxetanedione. The longer wavelength feature was proposed to arise from a structure containing a phenol group, possibly structure IV.

Minimizing the background emission is important to improving analytical detection. Hanaoka and co-workers,[26] in an analysis of the background emission in PO–CL, made use of differences in the decay rates of background and fluorophore emissions to optimize detection in HPLC. They found that the background emission consistently occurred with rates faster than those of fluorophore emission. By delaying the time to reach the flow cell of the detector, they could separate the two emissions and decrease the effect of the background emission.

8.1.3. Mechanistic Information Obtained from Photochemical Initiation

Recently, Milofsky and Birks demonstrated for the first time that the peroxy-oxalate reaction could be initiated photochemically in the presence of dissolved molecular oxygen and a hydrogen atom donor (HAD) such as isopropyl alcohol.[27] They found that the photo-initiated reaction and H_2O_2-initiated chemiluminescence exhibited similar characteristics, including (1) catalysis by imidazole and other bases, (2) greatly enhanced response to amino-PAHs as compared to the corresponding unsubstituted compounds, and (3) a linear response to fluorophore concentration.

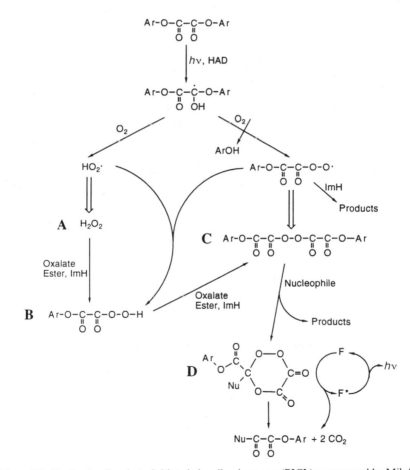

Scheme IV. Mechanism for photo-initiated chemiluminescence (PICL) as proposed by Milofsky and Birks.[13] (ImH = imidazole; Nu = nucleophile.)

In subsequent work, Milofsky and Birks studied the kinetics of photo-initiated chemiluminescence (PICL) to obtain new insight into the reaction mechanism.[13] Chemiluminescence was found to be produced in two pathways, summarized in Scheme IV: (1) a pathway that forms a high-energy intermediate directly from the oxalate ester, and (2) a pathway that forms H_2O_2 as an intermediate. Kinetics of the H_2O_2 pathway are described by

$$A \xrightarrow{r} B \xrightarrow{f} C$$

where A, B, and C represent pools of reactants, intermediates, and products, respectively. The chemiluminescence intensity is proportional to the concentration of species B, and both reaction steps are first-order, irreversible reactions. The integrated rate equation for the H_2O_2 pathway gives

$$I = \frac{Mr}{f - r}(e^{-ft} - e^{-rt})$$

where I is the intensity at time t; M is the maximum intensity; and r and f are first-order rise and fall rate constants, respectively. Both the rise and fall rate constants were found to vary linearly with TCPO concentration and to have the same complex functional dependence on imidazole (based catalyst) concentration. That the oxalate ester affects both the rise and the fall of the light emission was interpreted to mean that the intermediate B, formed in the reaction of the oxalate ester with H_2O_2, reacts with an additional oxalate ester molecule. This kinetics result led Milofsky and Birks to postulate the formation of peroxydioxalate (**C** of Scheme IV) as a key intermediate in both photo-initiated and H_2O_2-initiated peroxyoxalate chemiluminescence:

$$\begin{array}{c} \text{Ar—O—C—C—O—O—C—C—O—Ar} \\ \quad\;\; \| \;\; \| \qquad\qquad \| \;\; \| \\ \quad\;\; \text{O} \;\; \text{O} \qquad\qquad \text{O} \;\; \text{O} \end{array}$$

Analogous to the cyclization of peroxyoxalate (**IV → V**) to form an energy-transfer intermediate, Milofsky and Birks proposed the nucleophile-catalyzed cyclization of peroxydioxalate to form the cyclic intermediate **D** of Scheme IV:

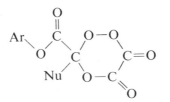

This species could form a charge transfer complex with the fluorophore, which would derive excitation energy from the decomposition of **D** into a nucleophile-substituted oxalate ester plus two CO_2 molecules. Formation of the six-membered ring structure, **D**, is expected to be greatly favored over the strained four-member rings of 1,2-dioxetanedione (**I**) and the proposed structure **IV** of Catherall et al. The decomposition of species **D** can also be expected to produce triplet states of carbon dioxide, possibly accounting for the background chemiluminescence observed by Mann and Grayeski.[25]

In the kinetics studies of the PICL reaction, oxalate ester concentrations were always in large excess over H_2O_2, making the reaction pseudo-first-order in H_2O_2. These reaction conditions made it possible to isolate for the first time the similarity of rise and fall dependence on oxalate ester concentration. Previous studies such as those of Catherall et al. and Alvarez et al. made use of reaction conditions in which the H_2O_2 concentration was comparable to or much higher than the concentration of the oxalate ester. Under these conditions, energy transfer intermediates other than **D** could be formed preferentially. Milofsky and Birks[13] proposed that peroxydioxalate (intermediate **C** in Scheme IV) could possibly serve as the partially stable intermediate **Z** in the Alvarez et al. Mechanism (Scheme III). A highly speculative but interesting possibility put forward by Milofsky and Birks is the formation of a cyclic peroxide following the reaction of H_2O_2 with peroxydioxalate:

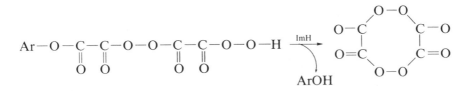

This cyclic peroxide could form a charge transfer complex with a fluorophore and transfer energy associated with its decomposition into four CO_2 molecules.

In the past few years, it has become clear that the PO–CL reaction is much more complicated than previously thought. The plausible reaction pathways are summarized in Scheme V and lead to six possible cyclic high-energy intermediates. Note that all of these pathways involve nucleophilic displacement of ArOH, which, by design, is a good leaving group. In light of recent work, the traditional pathway involving an energetically unfavorable

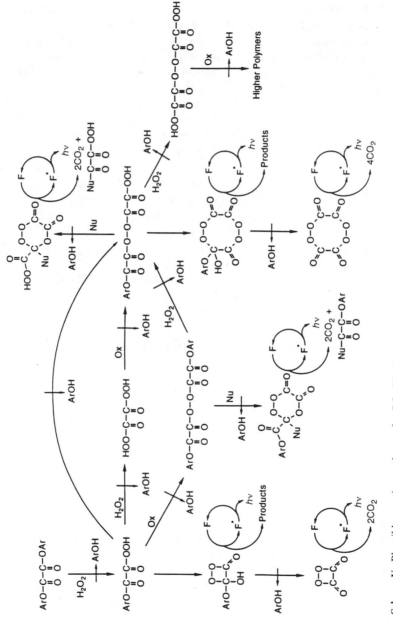

Scheme V. Plausible reaction pathways for PO–CL. Note that not all of these pathways need occur to meet the minimal requirements of Scheme III. (Ox = oxalate ester.)

220

four-membered ring cyclization (shown to the left and downward in Scheme V) is not necessary for explaining the multiple pathways observed for the reaction but cannot be ruled out at this time. Thus, the detailed mechanism of PO–CL remains an intriguing and unsolved mystery, largely because none of the reaction intermediates have ever been absolutely identified. Further characterization of the reaction mechanism and light-generating intermediates may, in principle, lead to the development of peroxyoxalate reaction detection having still greater sensitivity.

8.2. OPTIMIZATION OF PEROXYOXALATE CHEMILUMINESCENCE FOR ANALYTICAL APPLICATIONS

In this section, we address the design and optimization of a PO–CL detection method for flow injection analysis or HPLC. The design of an instrument based on PO–CL is relatively simple and comparable to setting up any postcolumn reaction to assist detection. The essential components for HPLC detection of fluorophores based on PO–CL are shown in Figure 8.2. In general, only a fraction of the emitted light occurs within the flow cell region of the detector. The "time-window" concept, illustrated in Figure 8.3, is useful for designing and optimizing detectors based on chemiluminescence.[24,28] The limit of detection for an analyte is strongly affected by the light intensity within the flow cell. The primary considerations for sensitivity and reproducibility using PO–CL are the reaction conditions, the delivery and mixing of the reagents, and design of the detector for maximum collection of light.

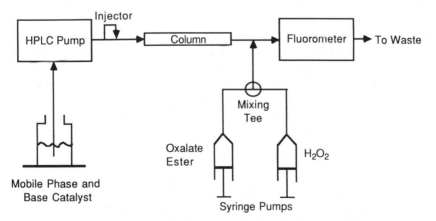

Figure 8.2. Schematic diagram of an apparatus for PO–CL detection of fluorophores.

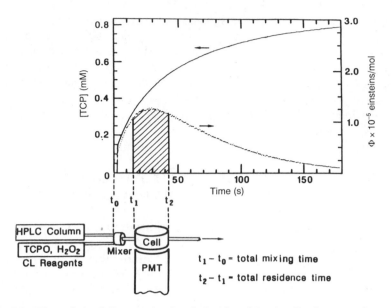

Figure 8.3. "Time-window" diagram showing the fraction of the chemiluminescence time profile observed by a detector in a flowing stream. (RP-HPLC = reversed-phase high-performance liquid chromatography.) Reprinted from Givens and Schowen,[24] with permission.

8.2.1. Optimization of Reaction Conditions

Optimizing PO–CL detection for HPLC begins with the selection of the oxalate ester. The most common oxalate esters for PO–CL detection are TCPO and DNPO. Mohan and Turro describe a synthesis for these oxalate esters,[29] but both are commercially available. The concentration of the ester typically ranges from 1 to 10 mM and is usually limited by its solubility in organic solvents such as acetonitrile and ethyl acetate. Several other oxalate esters, such as bis[2-(3,6,9-trioxadecyloxycarbonyl)-4-nitrophenyl]oxalate (TDPO) have been synthesized and evaluated for detection in HPLC.[30] TDPO is soluble in acetonitrile up to 1 M and is more stable in the presence of H_2O_2 than TCPO (31). The choice of oxalate ester depends on the application and results desired. For example, DNPO demonstrates faster reaction kinetics and greater CL intensities than TCPO. However, DNPO is less stable in solution and has a higher background signal than TCPO.

The concentration of H_2O_2 and base catalyst have the greatest effect on the PO–CL reaction. Typical concentrations of H_2O_2 for fluorophore detection are in the range of 30–500 mM. The base catalyst having the greatest effect on the CL intensity and kinetics is imidazole (28). The effect of increasing

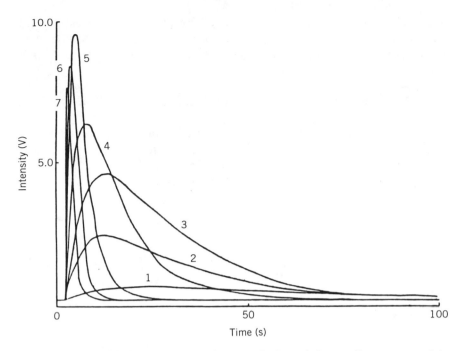

Figure 8.4. Effect of imidazole concentration on PO–CL emission profiles as measured by stopped flow. Reagent concentrations after mixing in flow stream: TCPO, 1.0 mM; H_2O_2, 5.0 mM; dansylalanine, 25 mM. Final imidazole concentrations after mixing in the flow stream: (1) 0.5; (2) 1.5; (3) 2.5; (4) 5; (5) 15; (6) 25; (7) 50 mM. Reprinted from Hanaoka et al.,[28] with permission of the American Chemical Society.

the imidazole concentration on the PO–CL reaction is shown in Figure 8.4. Imidazole concentrations in the range of 1.5–20 mM are commonly used in the mobile phase for FIA or HPLC. Since PO–CL is strongly affected by the base catalyst and peroxide concentrations, some experimental variation is required for optimization of the reaction.

Temperature, pH, and solvent composition also affect the CL intensity. PO–CL detection is favored at 30 °C.[28] Some efforts at temperature control should be made to maintain a reproducible CL signal. The PO–CL reaction is strongly affected by pH. The optimum pH for PO–CL ranges from 6 to 8 with sharp decreases in CL yield at low or high pH. DNPO has been used, however, when a lower pH is required.[32] Increasing the ratio of water in the solvent increases the background emission and lowers the chemiluminescence yield. Water appears to have a large effect on the rate of formation and lifetime of the various PO–CL intermediates. The low CL yield in solutions with high water content can be improved by changing the local solvent conditions after separation. For example, the postcolumn addition of reagents in organic solvents can be used to reduce the fraction of water in the detection cell.

8.2.2. Optimization of Delivery and Mixing of the PO–CL Reagents

Reproducible chemiluminescence requires pulseless flow for reagent delivery. Pulsed flow creates local concentration fluctuations in the reagents that introduce noise by modulating both the background and signal chemiluminescence. Nearly pulseless flow is often obtained with syringe pumps. Single-

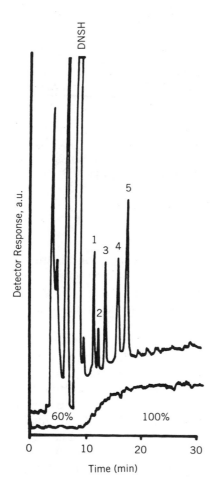

Figure 8.5. Analysis of a 5 L air sample (A) and field blank (B) taken at the University of Colorado campus, Boulder, on March 12, 1991, at 8:00 p.m. Samples collected on a silica cartridge and eluted directly onto a C_{18} column for detection by PICL: (1) formaldehyde; (2) acetaldehyde; (3) acetone; (4) benzaldehyde; (5) tolualdehyde. Reprinted from Nondek et al.,[35] with permission.

pump systems deliver a mixture of TCPO and H_2O_2. However, this approach results in a downward signal drift of about 1% per hour due to the decomposition of the oxalate ester.[32] The usual approach is to use a two-pump system to mix the reagents at the optimal time prior to the detector. An improvement in reagent delivery has been accomplished by using a dual-head, short-stroke pump.[33] This pump has become useful in flow injection analysis by reducing the volumes of solution segments to be mixed. Because concentration variations resulting from incomplete mixing are minimized, a more stable signal results.

Since syringe pumps are expensive, another option for one-pump reagent delivery is the use of photo-initiated chemiluminescence.[27] In this system, an ethyl acetate solution of TCPO (3 mM) and isopropyl alcohol (0.16 M) is pumped through a quartz reactor irradiated at 254 nm with a low-pressure mercury lamp. After irradiation, the photoproducts are mixed with the column eluate and react to generate chemiluminescence. Nondek et al.[34] used PICL detection of dansylated aldehydes and ketones in ambient air samples (Figure 8.5). Sensitivity comparable to fluorescence, ranging from five times poorer for the C_5 aldehyde to four times better for the C_9 aldehyde, was obtained. For all aldehydes, H_2O_2-initiated PO–CL detection was at least an order of magnitude more sensitive than fluorescence detection. The PICL system was simpler and easier to operate than H_2O_2-initiated detection and has potential for further improvements in sensitivity. A disadvantage of PICL relative to H_2O_2-initiated detection is greater quenching of the PICL reaction by water.

Efficient mixing is essential for reproducible results and high sensitivity. Incomplete mixing results in baseline noise and a variable analyte signal. A limitation to postcolumn mixing systems is the extracolumn band broadening caused by additional tubing and mixing devices. The rapid decay of chemiluminescence emission requires effective mixing devices having small internal volumes. Miyaguchi and co-workers have used a microbore rotating flow mixing device (Figure 8.6), which combines the column and reagent flows in a small diameter tube.[35] The combination of column eluent and reagents induces mixing and limits the dispersion of the peak profile. Other devices use low- or zero-dead-volume tees followed by coiled tubing to mix the reagents. The mixing coil is also useful for varying the time to reach the detector, thereby allowing the maximum CL intensity to occur in the detection window. The delay coil can be made from small inner diameter stainless steel tubing coiled to a diameter of 0.5–1.0 cm.

The problem of effectively mixing the PO–CL reagents with the column eluent is compounded by the need to react the ester, often dissolved in ethyl acetate, with the mixed aqueous solvents used in HPLC. Different ratios of water to organic solvent strongly affect the reaction conditions and may even

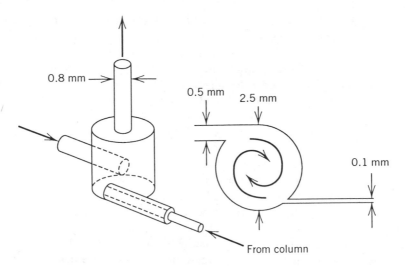

Figure 8.6. Rotating flow mixing device. Reprinted from Miyaguchi et al.,[36] with permission.

precipitate the oxalate ester in the mixing device.[32] This problem may often be solved by adding isopropanol or acetonitrile to the reagent solvent. The use of mobile phase gradients, necessary for complex separations, generates a noisy and sloping baseline during PO–CL detection. However, the variation in chemiluminescence intensity that causes baseline drift can be minimized by equally buffering both the column and the reagent mobile phases.[36] Since the baseline fluctuations are caused mainly by the background CL emission, an improvement in baseline stability also can be obtained by using a longer delay time between the mixing device and the detector, taking advantage of the observation that the background CL reaction decays much faster than the signal.[26]

8.2.3. Optimization of Light Detection

Commercial fluorescence detectors with the light source off and emission filters removed provide the most common method for PO–CL detection. The light emitted in the PO–CL reaction is detected by means of a photomultiplier tube (PMT). In the absence of scattered light from an excitation source, the PMT can be operated at higher voltages (higher gain) to obtain a higher signal-to-noise (S/N) ratio. Most fluorescence detectors are not designed for efficient collection of emitted light because of the need to discriminate against scattered light from the excitation source. A notable exception is the Kratos FS 970 detector, which makes use of an integrating sphere to collect approximately 50% of the emitted light. This detector was used in much of the early chemiluminescence detection work but unfortunately is no longer manufactured.

Detectors designed specifically for chemiluminescence detection are easily constructed. Although the Kratos and other fluorescence detectors have been useful for chemiluminescence detection, these detectors are nothing more than expensive PMT housings when used without a light source and emission filter. Gandelman and Birks[37] constructed a CL detector by placing the flow cell [coiled polytetrafluoroethylene (PTFE) tubing, 60 μL] at one focal point of an ellipsoidal mirror and the photocathode of a PMT at the other focal point.

The volume of the flow cell in the detector can be increased without significant loss of chromatographic resolution.[38] The bands of analyte separated in HPLC can be "chemically narrowed" because of the kinetic nature of the PO–CL reaction. Once the light-emitting reaction is complete, any additional cell volume is not sensed by the detector and therefore cannot contribute to band broadening. For example, with TCPO used as the oxalate ester, a sixfold increase in signal occurred when the cell volume was changed from 70 to 300 μL, with only a 10% increase in band broadening. Doubling the cell volume to 600 μL had no significant effect on the peak height. The use of DNPO for detection, which has faster reaction kinetics than TCPO, resulted in no change in the peak height or width using cell volumes in the range of 70–600 μL.[38] Faster kinetics and higher emission intensities do not always translate into lower detection limits. By forcing the reaction to be faster (e.g., by using DNPO in place of TCPO or by greatly increasing the catalyst concentration), the contribution of mixing noise is increased. As already discussed, special pumping systems must be used in order to take full advantage of the faster reaction kinetics of DNPO.

Several variables need to be considered when using PO–CL detection. The sensitivity of PO–CL detection is most strongly affected by the solvent, catalyst concentration, pH, and the delay time between reagent mixing and arrival at the detector. Optimizing the use of PO–CL for analytical detection involves fine-tuning the reaction conditions to efficiently generate light within the detection cell and designing the instrument to collect the largest possible amount of emitted light.

8.3. ANALYTICAL APPLICATIONS

Peroxyoxalate chemiluminescence detection is applicable to a wide variety of analytical problems. Although the most common application is the detection of fluorescent molecules, any of the other reagents, i.e., hydrogen peroxide, oxalate, or catalyst, may serve as the analyte. Sensitive detection has also been realized by enhancing the background emission resulting from the reaction of DNPO and H_2O_2 followed by quenching the standing CL signal by easily oxidized analytes. Discussed in this section are examples of how PO–CL has been used in various analytical schemes. For more complete information on applications, the reader is referred to two reviews published recently.[1,39]

8.3.1. Detection of Fluorescent Compounds

Polycyclic aromatic hydrocarbons (PAHs) represent the largest class of chemical carcinogens occurring in the environment. The amino- and nitro-derivatives of PAHs, also of environmental significance, are often even more mutagenic than the parent compounds. In a series of papers, Sigvardson and Birks[3,16,40] reported on PO–CL detection of PAHs, amino-PAHs, and nitro-PAHs in reversed-phase HPLC. An important finding of this work was the wide range of sensitivity exhibited by PO–CL detection toward different fluorescent compounds, attributable to the requirement of the formation of a charge transfer complex between the fluorophore and the high-energy intermediate. The amino-PAHs, which have low ionization potentials, were detected with as much as 10^5 greater sensitivity than the corresponding unsubstituted PAHs and with typically 1–2 orders of magnitude greater sensitivity than by fluorescence[16]. Limits of detection ranged from 90 fg for 1-aminofluoranthene to 8 pg for 2-aminofluorene. This selectivity toward amino-PAHs relative to other fluorescent compounds allows sensitive detection of these analytes in a complex matrix. In the chemiluminescence chromatogram shown in Figure 8.7, for example, amino-PAHs are selectively detected in an extract from shale oil. When fluorescence detection is used, the chromatogram is dominated by overlapping peaks of other fluorescent compounds in the complex matrix. By contrast, PO–CL detects only the amino-PAHs and results in a well-resolved chromatogram.

Nitro-PAHs are nonfluorescent and not detectable by either fluorescence or chemiluminescence. However, they may be quantitatively reduced on-line to the corresponding amino-PAHs by use of a short bed reactor of zinc particles, placed either before or after the analytical column, and detected with high sensitivity and selectivity by PO–CL.[40] As shown in Figure 8.8, the analytes may be determined with two sets of characteristic retention times, depending on whether precolumn or postcolumn reduction is used. Sigvardson and Birks applied this scheme to the detection of nitro-PAHs extracted from carbon black samples used in copying machines.[40]

An improvement in the detection of PAHs was recently discussed by Hayakawa et al.[34] They noted that using DNPO as the oxalate ester improved

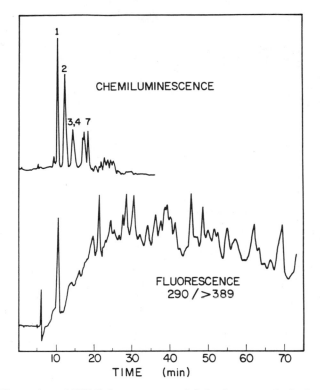

Figure 8.7. Comparison of HPLC chromatograms of shale oil extracts obtained using chemiluminescence and fluorescence detection. Fluorescence excitation at 290 nm with a 389 nm emission cutoff filter: (1) aminonaphthalenes; (2) methyl-aminonaphthalenes; (3) aminophenanthrenes; (4) aminoanthracenes; (7) aminopyrenes. Reprinted from Sigvardson et al.,[16] with permission from the American Chemical Society.

the detection limits by factors of 4–10 for certain PAHs. The oxalate ester and hydrogen peroxide were kept in separate reservoirs and mixed just prior to detection by a dual-head, short-stroke pump. The improved detection limits were related to the faster kinetics and greater chemiluminescence yield when DNPO is used.

8.3.2. Detection of Compounds Derivatized with a Fluorescent Tag

Since most compounds lack native fluorescence, derivatization plays an important role in the application of peroxyoxalate detection. The most commonly used derivatizing agent is the dansyl (DNS) tag, containing a reactive region attached to an aminonaphthalene ring. Being substituted

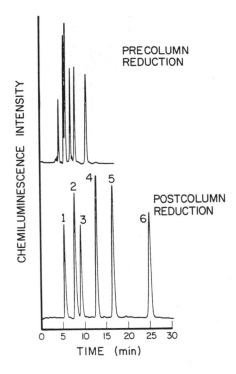

Figure 8.8. HPLC chromatogram of a mixture of six nitro-PAHs with PO-CL detection: (1) 1-nitronaphthalene; (2) 2-nitrofluorene; (3) 9-nitroanthracene; (4) 1-nitropyrene; (5) 6-nitrochrysene; (6) 3-nitroperylene (50–500 pg each). Reprinted from Sigvardson and Birks,[40] with permission.

amino-PAHs, dansyl derivatives have low ionization potentials and high fluorescence quantum yields, making them ideally suited to PO–CL detection. In the first application of PO–CL detection to HPLC, Kobayashi and Imai[2] reported the separation of dansylated amino acids, with detection limits in the femtomole range. Miyaguchi and colleagues later successfully used gradient elution for dansylated amines.[41] However, to minimize the sloping baseline and background noise, slow flow rates were used, causing the separation to take about 3 h. Hanaoka and Tanaka[42] improved this separation by using a longer delay time (20 s) between mixing and detection. The separation of dansylamino acids, shown in Figure 8.9, was accomplished much faster (30 min) and the baseline noise reduced.

8.3.3. Detection of Hydrogen Peroxide Produced in a Postcolumn Reaction

Hydrogen peroxide also may serve as the analyte in PO–CL detection. In fact, the first analytical use of PO–CL was reported by Seitz and co-workers for the detection of hydrogen peroxide.[43] In HPLC, hydrogen peroxide may

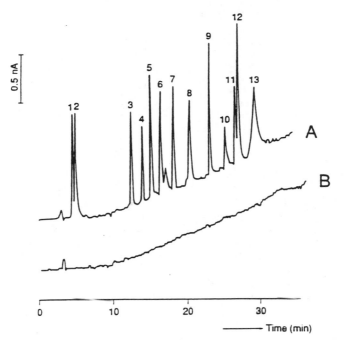

Figure 8.9. HPLC separation and detection of amino acids derivatized with dansyl chloride: (1) 82 fmol of Asp; (2) 85 fmol of Glu; (3) 77 fmol of Asn; (4) 62 fmol of Gln; (5) 92 fmol of Ser; (6) 80 fmol of Thr; (7) 75 fmol of Ala; (8) 67 fmol of Pro; (9) 80 fmol of Val; (10) 22 fmol of Lys; (11) 42 fmol of Ile; (12) 75 fmol of Leu; (13) 63 fmol of Phe. Gradient elution conditions: solution 1 consisted of 1.8 mmol of imidazole dissolved in 1000 mL water (pH 7.0); solution 2 consisted 1.8 mmol of imidazole dissolved of 400 mL of water and 600 mL of acetonitrile (pH 7.0). Linear gradient from 75% to 30% solution 1 in 35 min at 0.8 mL/min. (B) Baseline measured under the same gradient conditions as those in (A) without an LC column. Reprinted from Hanaoka and Tanaka,[42] with permission.

serve as the surrogate analyte, being generated in a postcolumn reaction. For example, Kamei et al. recently developed a detection system for polyamine compounds using PO–CL.[44] The determination of polyamines such as spermine, spermidine, putrescine, and cadaverine in biological fluids is an important monitor of the progress of cancer patients. The most common method uses postcolumn derivatization of the polyamine with o-phthalaldehyde (OPA) and fluorescence detection. However OPA reacts with a number of compounds in biological fluids, requiring strict separation of the polyamines or a sample pretreatment step. The system designed by Kamei and co-workers uses immobilized enzymes that react with polyamines to generate hydrogen peroxide. The acidic mobile phase (pH 2) was mixed postcolumn with a

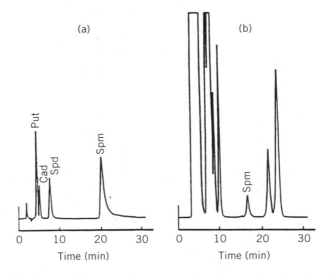

Figure 8.10. Chromatographic separation and detection of polyamines in a hydrolyzed urine sample: (a) H_2O_2 produced in a postcolumn enzyme reaction is detected by PO–CL; (b) detection by the *o*-phthalaldehyde (OPA) technique. Polyamines detected are putrescine (Put), cadaverine (Cad), spermidine (Spd), and spermine (Spm). Reprinted from Kamei et al.,[44] with permission from the American Chemical Society.

buffer solution containing imidazole to raise the pH to 7.5, and the separated polyamines were reacted with polyamine oxidase, immobilized on glass beads inside the enzyme column, to produce hydrogen peroxide. The flow stream containing hydrogen peroxide was then mixed with DNPO and 8-anilinonaphthalenesulfonate to produce chemiluminescence. A linear relationship between concentration and luminescent signal was observed for up to 500 pmol for cadaverine and up to 1 nmol for the other three polyamines. Results of the analysis of standard samples were comparable to those obtained by the OPA method. However, the chromatograms in Figure 8.10 show a clear advantage of PO–CL detection over the conventional OPA method. Owing to the substrate specificity of the enzymes, the polyamines can be detected with greater selectivity against the many OPA reactive compounds in urine.

Hydrogen peroxide may also be generated in a postcolumn photochemical reaction sensitized by quinones and detected by PO–CL.[45-47] A photochemical reaction scheme used by Poulsen and Birks[47] for the detection of quinones is illustrated in Figure 8.11. In photocatalytic chemiluminescence (PCCL) detection, advantage is taken of the catalytic nature of the reaction, which was demonstrated to produce several tens of H_2O_2 molecules for every quinone analyte molecule during an approximately 3 min residence time in

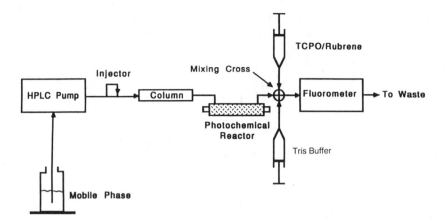

Figure 8.11. Schematic diagram of the photocatalytic chemiluminescence (PCCL) detection scheme used for the detection of quinones in HPLC. Reprinted from Poulsen and Birks,[47] with permission from the American Chemical Society.

the postcolumn photochemical reactor. The following simplified reaction scheme shows how a quinone, Q, can produce hydrogen peroxide in the photooxidation of the methanol component of the HPLC mobile phase:

$$Q + h\nu \rightarrow {}^1Q \rightarrow {}^3Q$$

$${}^3Q + CH_3OH \rightarrow QH^{\cdot} + {}^{\cdot}CH_2OH$$

$$QH^{\cdot} + O_2 \rightarrow HO_2^{\cdot} + Q$$

$$HO_2^{\cdot} + HO_2^{\cdot} \rightarrow H_2O_2 + O_2$$

The reaction is initiated by absorption of light by a quinone molecule to form an excited singlet state. This singlet state is rapidly converted to the lowest triplet state through intersystem crossing. The excited triplet-state molecule can abstract a hydrogen atom from methanol to form a semiquinone radical, $QH^{\cdot}$. In the presence of dissolved molecular oxygen, the hydrogen atom is rapidly donated to oxygen to form the hydroperoxyl radical, $HO_2^{\cdot}$, and regenerate the quinone. Disproportionation of two $HO_2^{\cdot}$ radicals forms H_2O_2. Because the quinone is regenerated, the reaction is catalytic and the quinone can repeatedly absorb light to form additional molecules of H_2O_2. The catalytic nature of the reaction greatly enhances the sensitivity and selectivity (since few classes of compounds can photochemically sensitize the formation of H_2O_2) of the technique. Applied to the detection of quinones, the method was found to be linear over at least 4 orders of magnitude and

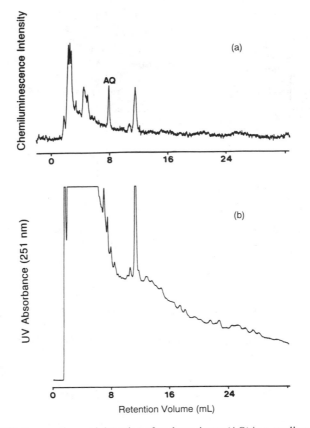

Figure 8.12. HPLC separation and detection of anthraquinone (AQ) in a cardboard extract: (a) PCCL detection; (b) UV absorbance detection at 251 nm. The anthraquinone peak is undetectable in the UV absorbance chromatogram. The retention volume axis in the PCCL chromatogram is offset by the volume of the photoreactor. Reprinted from Poulsen and Birks,[47] with permission from the American Chemical Society.

provided detection limits in the tens of femtomoles.[47] The selective detection of anthraquinone in a cardboard extract is illustrated in Figure 8.12.

The same photochemical reaction, in combination with PO-CL, has been used for the detection of hydrogen-atom donors.[48] In this case, acetonitrile, an extremely weak hydrogen atom donor, was used as the organic modifier in the HPLC mobile phase. A wide variety of pharmaceuticals that are either very poorly detected or not detected at all by ultraviolet (UV) absorption were detected at the low-nanogram level based on the photochemical production of H_2O_2. In this case the reaction is not catalytic, and the sensitivity is therefore much lower. The method represents, however, an approach to detection of compounds that have no chromophores and therefore cannot be detected by either UV absorbance or fluorescence.

8.3.4. Detection of Oxalates

An example of the versatility of PO–CL is the detection of oxalic acid in urine. A high amount of free oxalic acid in urine is an indicator of the formation of kidney stones. In an attempt to simplify the steps required to quantify oxalic acid, Albrecht and co-workers used PO–CL detection of oxalate in urine. Urine samples were collected and diluted with pH 1 hydrochloric acid. To each sample, $300\,\mu L$ of H_2O_2, $300\,\mu L$ of $25\,g/L$ bis(cyclohexyl)carbodiimide, and $50\,mg/L$ diphenylanthracene were added and the resulting signal detected in a luminometer. The chemiluminescence was measured for $4\,s$, and the area under the curve was used for determining the oxalic acid concentration. Despite the necessity of carrying out the reaction under unfavorable conditions of low pH and high water content, the sensitivity toward oxalic acid was adequate. Although the statistical difference between oxalate concentration in healthy persons and oxalate stone formers was found to be inadequate for diagnosis, the method represents a step in improving oxalate determination in biological samples.

8.3.5. Detection Based on Enhancement of the Background Signal

The enhancement of the background signal caused by certain analytes is a novel application of PO–CL detection. Capomacchia et al. used the enhancement of the background chemiluminescence resulting from the reaction of DNPO with H_2O_2 to detect the non-fluorophores urea and ouabain, a toxic glycoside.[49] Their flow-injection system mixed DNPO with H_2O_2 at low pH (pH 2–4). The background chemiluminescence signal was increased upon addition of urea and ouabain. The detection limits for urea and ouabain obtained by this method were 2 and 20 pmol, respectively. Both urea and ouabain contribute to the favorable formation of light-emitting intermediates in the absence of fluorophores. Further investigation into the effects of these compounds on the reaction pathways is needed.

8.3.6. Analysis by Quenched Peroxyoxalate Chemiluminescence

Easily oxidized nonfluorescers such as bromide, iodide, sulfite, nitrite, substituted anilines, and organosulfur compounds have been shown to quench the peroxyoxalate signal.[50] As a detection method, the presence of an analyte is monitored by the decrease in the standing signal produced by a mixture of an oxalate ester, hydrogen peroxide, and fluorophore. Apparently, easily oxidized compounds compete favorably for the reactive intermediate in PO–CL, causing a decrease in chemiluminescence yield. The selectivity of quenched PO-CL for easily oxidized compounds is useful in the analysis of biological samples. Van Zoonen et al. analyzed untreated and undiluted urine samples for methimazole and N-allylthiourea using the quenched signal of DNPO, H_2O_2, and 3-aminofluoranthene immobilized on silica.[50] Because only a small number of urine constituents induce quenching, the technique was highly selective, with detection limits in the low-nanogram range.

8.4. CONCLUSIONS AND FUTURE DIRECTIONS

Modern analytical chemistry continues to require the development of new techniques for the analysis of small amounts of samples, often ppb quantities and below in highly complex matrixes. Analysis by column liquid chromatography continues to be limited by detection capability. Most commercially available detectors rely on physical processes for detection, with fluorescence detection often used for sensitive analysis. It is hoped that the examples described in this chapter serve two purposes: (1) to illustrate the sensitivity and selectivity that can be achieved by using PO–CL detection, and (2) to foster new research, applications, and improvement of the technique.

The selectivity advantage of PO–CL detection has become increasingly useful in biomedical analysis. Specific fluorescent compounds can be selectively detected, along with the hydrogen peroxide generated from substrate-selective enzymes. The number of useful applications of various enzymatic reactions for sensitive assays of biological samples is expected to continue to grow.

The role of derivatization, important to many chemiluminescence applications, requires improvement. Several fluorescent tags generate artifact peaks and separation problems when used as precolumn derivatizing agents. Also, further research into labels that take advantage of the mechanistic features of the PO–CL reaction may further improve detection limits.

The use of photo-initiated peroxyoxalate chemiluminescence has received little attention in the literature since its discovery in 1990. The use of PICL greatly simplifies reagent delivery and by using laser excitation has the potential, at least, to provide much greater sensitivity than the H_2O_2-initiated reaction. Because the decay of the signal is determined by the reaction kinetics rather than the fluorescence lifetime, gating of the signal collection to remove scattered light from the laser is much simpler than for laser-induced fluorescence (LIF). PICL has the further advantage over laser-induced fluorescence of being much more selective. Simple calculations suggest that it may be possible to achieve single molecule detection with PICL, as has already been done for LIF.[51]

Isolation and positive identification of the intermediates involved in PO-CL would go a long way in elucidating the peroxyoxalate reaction mechanism, and improvements in analytical detection could also result from learning how to efficiently generate the key intermediates. For example, a problem with the PO–CL reaction is the low Φ_{CL} value, 0.1%, reported by Orlovic et al.[22] for the reaction of TCPO and H_2O_2 in aqueous solutions. This implies that more than 99.9% of the potential signal is lost to non-light generating reactions. Increased knowledge of the reaction mechanism may enable investigators to manipulate the chemistry so as to minimize the side reactions occurring in aqueous solution.

The advantages of PO–CL detection in HPLC, such as selectivity and reduced background noise, may also be applicable to capillary electrophoresis (CE). Capillary electrophoresis relies on the separation of molecules under the influence of an electric field. Although the separation efficiency is extremely high and analysis time very short, one of the main limitations of CE is the required detection of very small amounts of analyte. The sensitivity of chemiluminescence is well suited to low-volume, on-line detection because of the kinetic nature of the measurement. Chemiluminescence detection in CE has been demonstrated by using two other chemiluminescent systems. Dadoo et al. mixed potassium ferricyanide in a postcapillary reactor with luminol and H_2O_2 to generate chemiluminescence.[52] Ruberto and Grayeski used a similar postcapillary reaction scheme for the chemiluminescence detection of the reaction of acridinium esters and H_2O_2.[53] Both authors demonstrated attomol (10^{-18} mol) detection, but only in conditions of high pH, limiting the applicability of the method.

The low quantum yield of peroxyoxalate chemiluminescence in buffered aqueous solutions makes interfacing PO–CL detection with capillary electrophoresis difficult. Since the electrophoretic mobility is highly dependent on buffer properties such as conductivity and viscosity, addition of organic solvents to the buffer will greatly affect separation. In order to overcome this difficulty, PO–CL detection was applied to CE by combining electrophoresis with dynamic flow.[54] In this approach, the electrophoresis was stopped once the desired separation was complete within the column and a pump was used to elute the capillary into a postcolumn detection system. Although this demonstration of PO–CL detection with CE is novel, by not allowing all analytes to migrate through the entire column, much of the separation ability of electrophoresis is sacrificed. Clearly, it would be much more desirable to carry out detection without disturbing the electrophoretic separation. Again, a key is efficient detection in aqueous solvents.

Since its first application to chromatographic detection in 1980, the effectiveness and reliability of PO–CL detection has been demonstrated for many classes of compounds using a wide range of detection schemes. PO–CL detection offers a cost-effective means of detection at the attomole level with a much higher degree of selectivity than fluorescence detection. With recent advances in laser-induced fluorescence and the demonstration of single molecule detection, PO–CL now lags behind LIF in sensitivity but remains unequalled in selectivity. A hybrid technique making use of PICL has the potential for attaining the sensitivity and selectivity advantages of both LIF and PO–CL.

REFERENCES

1. K. Robards and P. J. Worsfold, *Anal. Chim. Acta* **266**, 147–173 (1992).
2. S. Kobayashi and K. Imai, *Anal. Chem.* **52**, 424 (1980).
3. K. W. Sigvardson and J. W. Birks, *Anal. Chem.* **55**, 432–435 (1983).
4. P. J. Kwakman, H. Schaik, U. A. T. Brinkman, and G. J. de Jong, *Analyst (London)* **116**, 1385–1391 (1991).
5. G. Mellbin and B. E. F. Smith, *J. Chromatogr.* **312**, 203 (1984).
6. E. A. Chandross, *Tetrahedron Lett.* **12**, 761–765 (1963).
7. L. J. Bollyky, R. H. Whitman, B. G. Roberts, and M. M. Rauhut, *J. Am. Chem. Soc.* **89**, 6523–6526 (1967).
8. M. M. Rauhut, B. G. Roberts, M. Loy, R. H. Whitman, A. V. Lannotta, A. M. Semsel, and R. A. Clarke, *J. Am. Chem. Soc.* **89**, 6515–6522 (1967).
9. M. M. Rauhut, *Acc. Chem. Res.* **2**, 80–87 (1969).
10. P. Lechtken and N. J. Turro, *Mol. Photochem.* **6**, 95–99 (1974).
11. C. L. R. Catherall, T. F. Palmer, and R. B. Cundall, *J. Chem. Soc., Faray Trans. 2* **80**, 823–834 (1984).
12. F. J. Alvarez, N. J. Parekh, B. Matuszewski, R. S. Givens, T. Higuchi, and R. L. Schowen, *J. Am. Chem. Soc.* **108**, 6435–6437 (1986).
13. R. E. Milofsky and J. W. Birks, *J. Am. Chem. Soc.* **113**, 9715–9723 (1991).
14. G. B. Schuster, *Acc. Chem. Res.* **12**, 366–373 (1979).
15. F. McCapra, *Prog. Org. Chem.* **8**, 231 (1973).
16. K. W. Sigvardson, J. M. Kennish, and J. W. Birks, *Anal. Chem.* **56**, 1096–1102 (1984).
17. K. Imai, K. Miyaguchi, and K. Honda, in *Bioluminescence and Chemiluminescence: Instruments and Applications* (K. Van Dyke, ed.), Vol. 2, pp. 65–76. CRC Press, Boca Raton, FL, 1985.
18. D. R. Maulding, R. A. Clarke, B.G. Roberts, and M. M. Rauhut, *J. Org. Chem.* **33**, 250–254 (1968).
19. J. J. DeCorpo, A. Baronavaski, M. V. McDowell, and F. E. Saalfeld, *J. Am. Chem. Soc.* **94**, 2879–2890 (1972).
20. M. M. Chang, T. Saji, and A. J. Bard, *J. Am. Chem. Soc.* **99**, 5399–5403 (1977).

21. C. L. Catherall, T. F. Palmer, and R. B. Cundall, *J. Chem. Soc., Faraday Trans.* 2 **80**, 837–849 (1984).

22. M. Orlovic, R. S. Givens, F. Alvarez, B. Matuszewski, and N. Parekh, *J. Org. Chem.* **54**, 3605 (1989).

23. H. P. Chokshi, R. G. Carlson, R. S. Givens, T. Kuwana, and R. L. Schowen, *Biomed. Chromatogr.* **4**, 96–99 (1990).

24. R. S. Givens and R. L. Schowen, in *Chemiluminescence and Photochemical Reaction Detection in Chromatography* (J. W. Birks, ed.), Chapter 5, pp. 125–147. VCH Publishers, New York, 1989.

25. B. Mann and M. L. Grayeski, *Anal. Chem.* **62**, 1532–1536 (1990).

26. N. Hanaoka, J. Tanaka, A. Nakamoto, and M. Takada, *Anal. Chem.* **63**, 2680–2685 (1991).

27. R. E. Milofsky and J. W. Birks, *Anal. Chem.* **62**, 1050–1055 (1990).

28. N. Hanaoka R. S. Givens, R. L. Schowen, and T. Kuwana, *Anal. Chem.* **60**, 2193–2197 (1988).

29. A. G. Mohan, and N. J. Turro, *J. Chem. Educ.* **51**, 528 (1974).

30. K. Nakashima, S. Akiyama, W. H. Wang, Y. Tsukamoto, and K. Imai, *Analyst* (*London*) **114**, 1413–1416 (1989).

31. K. Imai, Y. Matsunaga, Y. Tsukamoto, and A. Nishitani, *J. Chromatogr.* **400**, 169–176 (1987).

32. K. Imai and R. Weinberger, *Trends Anal. Chem.* **4**, 170–175 (1985).

33. K. Hayakawa, N. Imaizumi, and M. Miyazaki, *Biomed. Chromatogr.* **5**, 148–152 (1991).

34. L. Nondek, R. E. Milofsky, and J. W. Birks, *Chromatographia* **32**, 33–39 (1991).

35. K. Miyaguchi, K. Honda, and K. Imai, *J. Chromatogr.* **316**, 501 (1984).

36. N. Hanaoka, *J. Chromatogr.* **501**, 155–165 (1990).

37. M. S. Gandelman and J. W. Birks, *J. Chromatogr.* **242**, 21 (1982).

38. G. J. de Jong and U. A. T. Brinkman, *Chromatographia* **18**, 129–133 (1984).

39. P. J. M. Kwakman and U. A. T. Brinkman, *Anal. Chim. Acta* **266**, 175–192 (1992).

40. K. W. Sigvardson and J. W. Birks, *J. Chromatogr.* **316**, 507–518 (1984).

41. K. Miyaguchi, K. Honda, and K. Imai, *J. Chromatogr.* **303**, 173–176 (1984).

42. N. Hanaoka and H. Tanaka, *J. Chromatogr.* **606**, 129–132 (1992).

43. W. R. Seitz, D. C. Williams, and G. F. Huff, *Anal. Chem.* **48**, 1003–1006 (1976).

44. S. Kamei, A. Ohkubu, S. Saito, and S. Takagi, *Anal. Chem.* **61**, 1921–1924 (1989).

45. J. R. Poulsen, P. van Zoonen, C. Gooijer, N. H. Velthorst, and R. W. Frei, *Chromatographia* **21**, 587–595 (1986).

46. J. R. Poulsen, G. Gübitz, P. van Zoonen, C. Gooijer, N. H. Velthorst, and R. W. Frei, *J. Chromatogr.* **360**, 371–383 (1986).

47. J. R. Poulsen and J. W. Birks, *Anal. Chem.* **62**, 1242–1251 (1990).

48. I. Aichinger, G. Gübitz, and J. W. Birks, *J. Chromatogr.* **523**, 163–172 (1990).

49. R. N. J. A. C. Capomacchia, S. M. Hemingway, P. D'souza, W. Prapaitrakul, and A. Gingle, *Anal. Chim. Acta* **196**, 305–310 (1987).

50. P. van Zoonen, H. Bock, C. Gooijer, N. H. Velthorst, and R. W. Frei, *Anal. Chim. Acta* **200**, 131–141 (1987).

51. E. Brooks Shera, N. K. Seitzinger, W. M. Davis R. A. Keller, and S. A. Soper, *Chem. Phys. Lett.* **174**, 553–557 (1990).

52. R. Dadoo, L. A. Colon, and R. N. Zare, *J. High Resolution Chromatogr.* **15**, 133–135 (1992).

53. M. A. Ruberto and M. L. Grayeski, *Anal. Chem.* **64**, 2758–2762 (1992).

54. N. Wu and C. W. Huie, *J. Chromatogr.* **634**, 309–315 (1993).

CHAPTER

9

TALES OF A RELUCTANT INSTRUMENT MAKER*

JAMES E. LOVELOCK

Coombe Mill, St. Giles on the Heath, Launceton, Cornwall PL15 9RY, England

I grew up in England, mostly near London, in the 1920s and 1930s. It was a cultural environment confident in its belief that the only worthwhile occupation for the intelligent was the humanities. I can clearly remember being told as a child by my uncle Hugo Leakey that if I didn't learn Latin and Greek I would be denied the opportunity of reading and understanding the great philosophers in their own language, and consequently I would grow up gravely disadvantaged. He and most educators of the time really believed this, and children who showed promise were assiduously moved toward an education in the classics. You could not enter one of the better universities without first proving a proficiency in Latin at least.

In this environment, science came second place and was not for those with serious ambitions. Engineering was even lower in the pecking order and considered little more than a blue-collar occupation. This passion of the English for the humanities was not entirely foolish. At the time, England was trying to manage a large part of the world. Managers trained in the humanities were seen to have a better potential for dealing with colonial problems than chemists or physicists.

I was fortunate to escape this all-pervasive educational trap by being turned onto science at the early age of four. It was the result of a Christmas gift from my father of a box full of electrical odds and ends. What now would be a kit. Not only was it fun to experience hands on the result of energizing buzzer lights and so on, but I began to wonder why it was necessary to pass electricity along two wires. Why was one not enough like water or gas. No one in the family could answer my simple question, so I was forced to think. From then on the world became for me a place in which to do

*Pittcon lecture, March 1994, Chicago.

Selective Detectors, edited by Robert E. Sievers. Chemical Analysis Series, Vol. 131.
ISBN 0-471-01343-9 © 1995 John Wiley & Sons, Inc.

experiments and make measurements. My cultural environment was set by writers like Jules Verne and H. G. Wells. By the time I was eight years old I was on to the hard stuff like Wade's textbook on organic chemistry and Eddington's books on astronomy. It was several years before I was exposed to the pointless and dull science taught at school. I was crystallized as the kind of scientist I was to become.

The wonderful thing about science that makes it different from all other intellectual activity is the possibility of testing one's speculations by experiment. Furthermore, the act of testing often required the inventing and building of an instrument, however crude, to do it. I recall making a simple vane anemometer to measure the speed of the wind. I calibrated it by holding it from the window of a moving train. The train speed was easily found by timing the travel between mile markers along the track. In short, by the time I reached puberty I saw my life's work as the reduction of science fiction to practice.

Of course I knew that to achieve this ambition I would have to submit to a long tedious and largely pointless education. By the 1930s there was no chance of diving straight in as a apprentice. Michael Faraday did it in the previous century, but it was no longer possible. To become a scientist without becoming at least a university graduate was as unthinkable as practicing medicine without an M.D. It was with thoughts like these that my headmaster told me when I left school at 18, "You are a fool, Lovelock, to take up science as a career. It is only for those with genius or those with ample wealth. You have neither of these, and when you come to get married you will realise that you have chosen a way of life that will never support a wife and family." These harsh words were well meant at the time, which was near the end of the 1930s' Great Depression. Graduate chemists were being hired by some companies at zero salary on the grounds that they were gaining industrial experience for free. The reason I am here now talking to you about it and not spending my last years in the poorhouse is partly because I was single-minded and obstinate in my wish to become a scientist, but mainly because I had amazing good luck.

My first stroke of luck was to find employment with a firm of consulting chemists in London immediately from school. This firm, called Murray Bull and Spencer, was run by a pair of academics from London University. Their field was organic chemistry and photochemistry, and they specialized in solving problems for the photographic industry and indeed for anyone who sought their services. They took on schoolboys and trained them to be lab technicians, but warned that there was no future in the firm for them. Humphrey Murray, my boss, was quite open about this, but redeemed what would now seem a heartless attitude by paying for his apprentices to attend London University as evening class students. More important for me than the formal education was the personal insistence from my boss that the analyses I did were for real and not student experiments to emphasise a point of teaching. This way, unlike a full-time university student, I grew to regard accuracy in measurement as almost sacred. As Humphrey was always saying, peoples lives and jobs can depend on the right answer, and the sloppy results—so normal in the university—became for me an abomination.

The hands on experience I gathered then was a priceless gift that has served throughout my life. One of the firm's customers made gelatin of photographic quality for Kodak. One day they were in serious trouble, for the photographic emulsions made with their gelatin suddenly became only weakly sensitive to light. Humphrey suspected a lack of sulfide in the gelatin and sent me to the firm in the company of a senior technician to have a look at the process and find out what had gone wrong. It was quite an experience. I had never realised that gelatin came from the indigestible leftovers of the meat industry. Hooves and skin were boiled in vast to make the thick soup that was allowed to gel and then dried as sheets of gelatin. We asked the foreman if he could think of anything that had changed in the process during the past few weeks. He replied, "Nothing has changed; we are doing exactly everything the same as before." My colleague noticed an empty rusty bucket next to one of the vats and asked what it was. The foreman replied, "That is the old bucket from which we add the hydrogen peroxide to clear the gelatin at the final stage of cooking. We bought a new one last week; here it is." Light began to dawn: we knew that the lack of sulfide could have come from an excess of oxidant, and here the new bucket was visibly larger in size than the old one. We soon found that the new bucket was twice the volume of the old one, and the problem was solved. This small experience made real for me the academic fact that volume increases as the cube of the linear dimensions. The foreman buying a new bucket from memory thought the one-third increase in diameter and depth of small consequence. This is the kind of experience that universities can rarely ever supply.

I would have stayed happily at Murray Bull and Spencers until I had my B.Sc. degree from London University, but it was then 1939 and Hitler had other ambitions. When war was declared in September 1939 all of the London colleges closed and students were sent to towns thought to be less under threat from bombing. I had the chance to enrol for the second two years of my degree as a full-time student at Manchester University. Here, the professor of chemistry was Alexander Todd, probably the best chemist in the United Kingdom. He was a young man in his thirties, full of idealism, and a fine professor to his students. He also treated us like apprentices and would spend a great deal of time talking to us personally as we worked in the laboratory. I had a nasty surprise though only one month after starting at Manchester. I was called to the professor's office. Todd looked at me sternly and straight in the eye and said, "I'll not have cheating in my department, Lovelock." I was amazed and confused, but he went on to say. "Look at the results of the gravimetric analysis you turned in last week. You have reported exactly the concentrations of bromide ion in the two solutions you analyzed. You may not know it, but students almost never get the right answer to a gravimetric analysis, and certainly never twice running. There is only one possible explanation; you must have looked up the answers in the class book that the demonstrator foolishly left in the lab." By then he was in full flood, telling me of my wickedness and the consequences for my career if I did not mend my evil ways. After he ran out of breath I said, "But I didn't." It took nearly 30 minutes to convince him that I was, after two years at the London consultants, a pro at this analysis and it was for me a routine task and one I expected to get right. The exchange left us both wondering a little what university training was really about.

I completed my studies at Manchester in June 1941, and by that time the war completely pervaded all aspects of life in England. Everyone was conscripted in one way or another to serve. Science was suddenly an accomplishment of great value to the nation, and our skills were appreciated as never before or again. It was my good fortune to be sent to one of the foremost scientific institutions in the United Kingdom, the National Institute for Medical Research in London. My senior colleagues there were all a little disconsolate at having to abandon their favorite academic projects in favor of providing answers to urgent war-related problems. Among the problems I was involved in were the prevention of the spread of respiratory infection among aircrew. A common cold, just a minor matter on the ground, could be an agonizing disability in an unpressurized plane at 30,000 ft. Then there was the fear of the consequences of the dense crowding of people in London's underground bomb shelters; so bad was the air that in one shelter the oxygen fell to 13%. We thought that a disease like influenza could be devastating under such conditions. Then we were asked to find ways to protect combatants from heat radiation. The heat radiation flux from the large fires inevitable in war are enough to produce on bare skin a second-degree burn at a distance of 100 yd from the source. We soon found that comparatively thin layers of fabric, particularly wool, were amazingly effective. A great deal of the equipment used for the wartime measurements had to be invented on the run and made from anything that came to hand. I well remember my boss—a lovely courteous man, Robert Bourdillon, a physician and World War I pilot—asking me if I could produce by the next day a device that would register the flux of heat radiation so as to indicate the probability of a first-, second-, or third-degree burn. It would have to be robust and usable under battlefield conditions by nonscientists. I cast aside thoughts of thermocouples and sensitive meters. I didn't even consider electronics: the firebottle amplifiers were too delicate, and the power they needed painful to wartime batteries. I wondered if crystals of substances with melting points in the range 50–100 °C spread on black paper would do the trick. It was not too bad but messy. Then I wondered if paper painted with the paint specially made to sense mustard gas would form a better background for the crystals. The gas-detecting paint was simply a green paint with particles of a red dye dispersed in it that was soluble in mustard gas. A droplet of mustard dissolved the dye and revealed itself as a red dot on a green background. My idea was that crystals chosen with solvent properties similar to mustard gas would have the same effect when they melted. It worked after a fashion, but then luck was with me once more, for I noticed that plain paper coated with the gas-detecting paint turned pink at the radiation flux that would give a first-degree burn dark red for a second, and yellow for a third-degree burn. It was a simple reproducible answer that could even be made on the spot where it was needed. No modern war could run without paper, and gas-detector paint was standard issue.

It was enjoyable and fulfilling work, but I believed my superiors who said that this was not real science, just ad hoc invention.

Wars are by no means all bad. Science has always flourished in times of war. I remember telling protesting students in the 1970s who would not do research funded by the military, "You are denying a respectable tradition going back to Archimedes and including such greats as Isaac Newton. His work on the Laws of Motion was supported by a ballistics contract from the British navy, as was Ernst Mach's similar research supported by the German army." So it was in the 1940s in London. Science was improved by the war and I benefited personally from the democratic obligation of all scientists at the institute to take their turn as firewatchers. We had to spend a night each week on duty. If there was an air raid, we had to put out fires caused by the numerous small incendiary bombs the fell like rain from the sky. Under these conditions, and when larger bombs were falling nearby, I found eminent older scientists suddenly suffered a kind of mental catharsis, an irresistible desire to pass on all they knew to whoever young was nearby. From this process I received a concentrated distillation of the essential knowledge of some of our finest scientists. What a priceless gift for a young apprentice scientist.

At the end of World War II my senior colleagues heaved a sigh of relief and returned to their long-term researches. I was now thrown into an environment where invention was low class and where pure science and publication was what counted. I soon found that I did not like it. To make matters worse, I was shipped to an outstation of the Medical Research Council at Salisbury: the common cold research unit. Here human volunteers were used as experimental animals in the search for the cause and perhaps the cure of the common cold, still one of the most difficult problems in medical research. Being practical, I think I did some good by turning my thoughts to how the infection was spread and succeeding with my colleagues in showing that colds were transferred only by close contact, not by fine airborne particles as was thought at the time. We even showed that the linen handkerchiefs then in use to mop up the profuse secretion of a cold were also ideal agents for transferring infective secretion to others and to the environment. Perhaps we helped establish the far more sanitary throw away paper tissues now used.

But my heart was not really in this kind of science. I responded by testing the common belief that colds are caused by exposure to a draft of cold air. To test it I needed an anemometer sensitive enough to measure the very low air velocities associated with a draft. None existed, so invention was required. It occurred to me that maybe an ionization anemometer would serve. Ions move slowly though air in a weak electric field, so it ought to be easy to observe their deflection. I made an ion anemometer by scraping the radioactive dial paint from a number of surplus aircraft altimeters. The radiation levels were astonishingly high enough to turn the hair of a health safety inspector white with concern. There were microcuries of radium on each dial—under wartime conditions it was considered that the risk to the health of aircraft crew from this radiation was trivial compared with the other hazards they faced. The ion anemometer was a cage made of three great circles of rigid wire enclosing concentrically a ball coated with radium from the dials. The alpha particles from the radium gave an ion current of several nanoamperes with a field of $10\,V/cm$. The smallest air movement reduced the current. We found that we could detect air movements as low as a foot per minute. More than sensitive enough for the measurement of drafts, it helped to show that exposure to cold or to drafts had nothing to do with catching a cold. Its only drawback, apart from the use of radioactivity, was that it was sensitive also to halocarbons like the CFCs (chlorofluorocarbons) and to cigarette smoke. These two anomalies of the anemometer were later to become the basis of the ubiquitous ionization smoke detectors and the ECD (electron capture detector).

In 1951 I was called back to the parent institute in London. The director, Sir Charles Harington, told me that my new task was to bring, as he put it, scientific credibility to the amazing discoveries of the biologists Chris Polge, Audrey Smith, and Alan Parkes: that cells and tissues could be frozen and kept viable in solutions of glycerol. Many good scientists in those days regarded biology as a second-class science or even just a craft, and the director was I think one of these. He said, "They are doing really important work, but I doubt if they know one end of a thermometer from the other. Your job is to see that they don't discredit the institute by doing bad science."

Unfortunately for me someone overheard his remarks and it took me a year before the biologists would trust me. Later I was able to collaborate with them, and we succeeded in freezing and reanimating a whole animal, a hamster. There were many tricks in doing this successfully, and one of them was the need to warm the heart and set it beating before the skin and the rest of the body warmed enough to consume oxygen. They did it by placing a hot metal spoon on the hamster's chest over its heart. It worked, but at the price of severe chest burns. Animal experimenters never seem to worry about this kind of event, but I found it sickening and hastened to invent a means to warm the animals internally without burning them. I made at home a simple diathermy apparatus that generated about 50 W of RF energy, and with this Audrey Smith was then able to reanimate her hamsters kindly. A little later I borrowed from the navy a 1 kW CW (continuous wave) magnetron. With this we were able to heat small tissue sample from the deep-frozen state at amazing speeds and entirely without harm. It was also used to reanimate frozen animals. I used it occasionally to cook potatoes for my lunch. I have often wondered if this was the first microwave oven.

It was the freezing work that led me to gas chromatography and a long association with the scientists of the instrument industry here in the United States. I found that the damage caused by freezing living cells and tissue came not as expected from the disruptive action of ice crystals but instead from the concentration of salts and other solutes as water froze out as the pure substance ice. The vulnerable part of cells are their membranes. They are made of lipids, phosphatides, and cholesterol and held together by van der Waals forces, not by covalent bonds. Strong salt caused the membranes to dissolve. Now the specific sensitivity of cells seemed to depend upon the fatty acid composition of the lipids, the more unsaturated the less easily damaged by freezing. It so happened that Archer Martin and Tony James, inventors of gas chromatography, were working in a lab close by, and knowing their early GC had cut its teeth on the analysis of fatty acids I asked them if they could analyze the fatty acids from my membrane lipids. They were enthusiastic to try until I showed the quantity I had for analysis. It was a few micrograms. Martin said, "Sorry, no way to do it. You will have to go back and bring us milligrams of the material; that is the least we can analyze." "Alternatively," they said, "you could invent a more sensitive detector than our gas density balance."

One does not hear much of the gas density balance these days as a detector, yet in 1950 it was as much an example of Martin's genius as was the invention of GC itself. The GDBs used in Martin's lab were exquisitely crafted three-dimensional plumbing within a block of solid copper. They were as sensitive as present-day TCDs (thermal conductivity detectors) and of course were absolute detectors and required no calibration. The first GCs seem very odd compared with the compact desktop products of today. The columns were straight 5 mm diameter glass tubes and always mounted vertically. They were rarely more than 4 ft long and were heated by enclosure within a metal jacket holding some fluid boiling at the desired temperature. Glycol boiling at 190 °C was commonly used. Samples were literally put on the columns. The pressure was let down, the top opened, and then the sample added by hand with a microliter pipette. A single fatty acid methyl ester analysis on a 4 ft apiezon column took several hours to run. Among the problems we faced was the impatience of chemists in the institute to check their synthesis by the world's only GC. They would steal in unannounced and put a sample on during a fatty acid run. We would find, in the gap between the long slow peaks of stearic and arachidic acid methyl esters, a small forest of sharp peaks, not a natural product but the chemist Gilbert Beavan's biphenyls. The institute's fire regulations precluded the use of hydrogen as a carrier, which would have increased sensitivity and speed of analysis. Helium in those days was far too expensive for us to use. It was against this background that I tried to develop a more sensitive detector. The story has been told elsewhere, but again it was entirely a matter of construction by hand. This time I did not have to take the risks of handling radium or polonium as were used with the anemometer. The first radioactive sources for detectors were pieces of silver foil containing ^{90}Sr, which was quite tricky to handle since the strong beta radiation from 200 MBq would quickly exceed safety limits for bare skin. But once inside the detectors they were, as now, safe and secure. Back then Teflon was not easy to get, and we used a spark plug as the insulator that held the anode. During those productive years at the Mill Hill Institute most of what is now commonplace in GC was tried or discussed, including liquid chromatography on columns or paper, supercritical fluid chromatography, and all manner of ionization detectors. It is easier to list what we did not do, which was invent the flame ionization detector (FID), the capillary column, and temperature programming.

Had our philosophy been in tune with the present-day politics of market forces we could probably have enriched the institute sufficiently to keep it running forever. As it was, the United Kingdom was steeped in socialism and we gave our advice and patentable ideas away, almost always for no return. This mattered little to the individual scientists—we were all well paid. But it was foolish as a national policy not to recover something from our contributions. Among the many visitors were those who brought as much or more than they took—like Dennis Desty, whose gift of a copper squalane coated capillary enabled me to invent and describe detectors with small internal volumes like the photoionization detector, the argon and helium detectors, and the ECD. Others whose gifts were prized were Ray Scott,

Tedd Adlard, and, among the visitors from the United States, Sandy Lipsky, Al Zlatkis, and Evan Horning.

In 1961 I received a letter from the director of space flight operations of NASA that was to change my life. I was invited to be an experimenter on the forthcoming *Surveyor* missions to the Moon whose object was to analyze and examine the lunar regolith ahead of the landing of the astronauts. How could anyone whose life plan was to reduce scifi to practice refuse such an offer? It led to the fulfilling and rich field of science opened up by the view of the Earth and the planets from above. It led me to what I like to think is the most significant invention of my life, namely, the theory of the Earth as a self-regulating system—what I like to call Gaia. I shall always be indebted to NASA for that invitation. It led me to become an independent practitioner of science and to a great number of friends and colleagues in America. I am especially indebted to the Hewlett Packard Company for their unstinted support over the last 32 years and for the wonderful friendships that developed as a consequence.

Working as an independent practitioner of science, my view from outside sees it fragmented, with each branch subdivided into ever smaller twigs of specialty. This is why so little is known about the Earth, about the way the oceans, atmosphere, and ecosystems work. To understand the Earth requires a top-down view, one that the myopic vision of the specialist can never provide. Science is now so specialized that scientists no longer can speak to one another. I wonder how many chemists would now be able to do hands on the analyses now done by the complex semi-intelligent instruments on show here? I wonder how many of them know how they work or could fix them if they went wrong. And how many wonder if they are being fooled by them?

Let me finish by showing you the kind of modern high technology that appeals to me. Here is a test paper used in clinical chemistry. A physician can dip this in a urine sample and in a few second know with sufficient accuracy the bacterial content, pH, sugar, ketones, protein, and the presence of traces of blood. It is the direct descendant of litmus paper—and a delightful example of the way science can get ahead and stay simple.

INDEX

Acetone, oxygen determination in, 110

Acids, redox chemiluminescence detection of, 92

Aircraft, measurements taken from, 45

Air pollution, gas chromatography/atomic emission studies on, 156

Alcohols, redox chemiluminescence detection of, 92

Alkyltin, chromatography/plasma mass spectrometry detection of, 183

Alternating current plasma (ACP) detector, as atomic plasma emission chromatographic detector, 144, 147

Aluminum, gas chromatography/atomic emission determination of, 149, 153, 163, 164

Amino acids:
  detection by peroxyoxalate chemiluminescence, 230
  sulfur-containing, aqueous phase conversion detection of, 25, 28

Ammonia, thermal conversion to nitric oxide for detection, 86–88

tert-Amyl methyl ether, oxygen determination in, 110

Antimony, gas chromatography/atomic emission detection of, 149, 151

Aqueous phase oxidation, for sulfur monoxide formation, 22–30

Aqueous solutions. See also Seawater; Water
  nitrite and nitrate ion detection in, 75–80

Argon chloride interference elimination, in chromatographic detection by plasma mass spectrometry, 189–190

Arsenic:
  chromatography/plasma mass spectrometry detection of, 178, 179, 180, 184–185, 186–187, 189–190
  gas chromatography/atomic emission detection of, 149, 151, 153, 156, 159

speciation of, 172, 180, 184–185, 189–190

Arsenobetaine, 172

Arylthiophenes, gas chromatographic/atomic emission studies on, 156

Atmosphere:
  nitric oxide determination in, 85–86
  nitrogen compound determination in, 85–91
  sulfur compound detection in, 1, 67, 156
  volatile organic compounds in, redox chemiluminescence detection of, 93

Atmospheric particles, nitrate ion detection in, 77

Atomic emission detector (AED):
  advantages of, 37
  comparison with other detectors, 66, 67
  for sulfur compounds, 37
  use in sulfur-selective detection, 12

Atomic plasma emission spectroscopy, element-selective gas chromatographic detection by, 143–169

Auranofin:
  chromatography/plasma mass spectrometry detection of, 178, 181–182
  as gold-containing drug, 173

Automotive emissions, oxygenated hydrocarbon determination in, 114

Bacon (fried), nitrosamine detection in, 82

Benzodiazepin, deuterium determination in, 153

Benzothiophene(s):
  detection by supercritical fluid chromatography, 65
  response factor relative to thiophene, 52

Bergamot, oxygen-sensitive detector studies on, 118

Biological samples, trace element determination in, 172–174

251

Biscyclopentadienyliron, as reagent for hydrogen atmosphere flame ionization detector, 130

Blood:
    arsenic detection in, 172
    gold drug detection in, 187

Blood plasma:
    carbon and fluorine determination in, 152, 153
    chromium determination in, 163
    organic nitrate detection in, 82

Boron, gas chromatography/atomic emission detection of, 149

Bovine serum albumin, aqueous phase conversion detection of, 27

Bromine:
    chromatography/plasma mass spectrometry detection of, 179, 191, 193, 194, 195, 199
    gas chromatography/atomic emission detection of, 147, 148, 149, 151

Bulk property detectors, use in gas chromatography, 143

1,3-Butadiene, cracking capacity of, 108

1-Butanol, oxygen determination in, 109

2-Butanol, oxygen determination in, 110

Cadaverine, detection in cancer patient fluids, 231

Cadmium:
    chromatography/plasma mass spectrometry detection of, 178, 186
    in foods, 172
    on-line preconcentration of, 188
    speciation of, 172, 186

Cancer patients, polyamine detection in fluids of, 231–232

Capacitively coupled plasma (CCP), as atomic plasma emission chromatographic detector, 144, 147

Carbon, gas chromatography/atomic emission determination of, 147, 148, 149, 151, 152, 153, 154, 160, 164, 166

Carbon disulfide, redox chemiluminescence detection of, 92

Carbon monoxide:
    formation in oxygen-sensitive detector, 101
    redox chemiluminescence detection of, 92

Catalytic thermionic conversion, sulfur monoxide formation by, 19–22

Cheese, nitrosamine detection in, 82

Chemiluminescence:
    applications to sulfur-selective detection in chromatography, 1–33
    definition of, 71
    nitric oxide/ozone detection method using, 71–97
    peroxyoxalate type. See Peroxyoxalate chemiluminescence

Chemiluminescence detectors, 40–42

Chicken, zinc detection in, 188

Chlorine:
    chromatography/plasma mass spectrometry detection of, 179, 191, 192, 193, 194, 195, 199
    gas chromatography/atomic emission detection of, 147, 148, 149, 154

Chromatography, sulfur-selective chemiluminescence detection in, 1–33
    flameless method, 35

Chromatography-atomic plasma emission spectroscopy (C/PES), applications of, 144

Chromium:
    chromatography/plasma mass spectrometry detection of, 111, 178, 185, 190, 199
    gas chromatography/atomic emission determination of, 149, 163, 165
    "paradox" of, 173
    speciation of, 172–173

City air, sulfur detection in, 156

Coal:
    distillate, element detection in, 155
    sulfur determination in, 147
    volatile aromatic hydrocarbons (PAHs) determination in, 115

Cobalt, gas chromatography/atomic emission determination of, 164, 165

Cobalt protoporphyrins, chromatography/plasma mass spectrometry detection of, 179

Collapsing cone mechanism, in hydrogen atmosphere flame ionization detector, 128, 129, 142

Copper:
    chromatography/plasma mass spectrometry determination of, 186, 187
    gas chromatographic/atomic emission determination of, 164, 165

on-line preconcentration of, 188

Cracking reactor, of oxygen-sensitive
detector, 103–105

Criegee-like intermediate, from ozone
reaction with C—S bond, 13

Crude oil:
element determination in, 155
metalloporphyrin determination in, 165
oxygenated hydrocarbon determination
in, 115, 119
sulfur compounds in, 36, 59

Cutting fluids, nitrosamine detection in, 83

Cyalume "light sticks," peroxyoxalate
chemiluminescence as basis of, 209

Dansyl tag, use in peroxyoxalate
chemiluminescence detection, 229–230

Detergents, use in micellar liquid
chromatography, 182

Deuterium, gas chromatography/atomic
emission detection of, 149, 151, 160

Di-t-butyl disulfide, response factor relative
to thiophene, 52

Diesel fuel, sulfur limits in, 59

Diethyl sulfide, response factor relative to
thiophene, 52

Dimethylarsinic acid (DMA), toxicity of, 172

Dimethyl sulfide, response factor relative
to thiophene, 52

Dimethyl sulfoxide (DMSO),
chemiluminescence response to, 19, 23

1,2-Dioxetanedione, structure of, 211

Dioxetanes, structure of, 211

Dioxetanones, structure of, 211

Diphenyl sulfide, response factor relative
to thiophene, 52

Direct current argon plasma (DCP)
detector, as atomic plasma emission
chromatographic detector, 144,
147–148

Direct injection nebulization, in
chromatographic detection by
plasma mass spectrometry, 189

Disulfides, detection by aqueous phase
conversion, 25

Dogfish muscle, arsenic determination in, 179

Electrolytic conductivity (Hall) detector, 37
comparison with other detectors, 66–67

disadvantages of, 42
use in element-selective gas
chromatography, 144, 153

Electron capture detector:
applications of, 92
use for halogen detection, 153

Environmental analyses:
using gas chromatography-atomic
emission, 151–152
using peroxyoxalate chemiluminescence,
228

Enzymes, peroxyoxalate chemiluminescence
studies on, 236

Ethanol, oxygen determination in, 109, 110

Ethyl tert-butyl ether, oxygen
determination in, 109

2-Ethylthiophane, response factor relative
to thiophene, 52

Ferrous ion, reaction with nitric acid, 94

Ferrous sulfate, use in nitrogen dioxide
reduction, 88

Fish, nitrosamine detection in, 82

Flame ionization detector (FID), 2, 49, 51
metal selectivity in. See Hydrogen
atmosphere flame ionization detector
detector response, 127–129
instrumentation, 129–131
for oxygen. See Oxygen-sensitive detector
(O-FID)
sulfur monoxide formation in, 38
use with oxygen-sensitive detector, 118,
120, 121
use with supercritical fluid
chromatography, 198–199

Flameless sulfur chemiluminescence detection:
applications, 59–67
block diagram for, 46
comparison with other methods, 66, 67
detector for, 19, 30, 43, 44
linearity, 54
operation and performance, 45–59
precision, 54–57
relative response factors, 52–54
selectivity, 51–54
sensitivity, 48–51
flame sulfur chemiluminescence
detection compared with, 43–45,
57
gas chromatography use with, 59–63

Flameless sulfur (*cont.*)
    liquid chromatography use with, 66
    supercritical fluid chromatography use
        with, 64–66
Flame photometric detector (FPD):
    advantages of, 37
    chemiluminescence efficiency, 5
    chemiluminescence mechanism, 3–5
    comparison with other detectors, 66
    design of, 5
    disadvantages of, 42
    dual-flame type, 5–7
    liquid chromatography use with, 66
    response characteristics and applications
        of, 7–8
    single-flame type, 5, 6
    for sulfur compounds, 37
    use with element-selective gas
        chromatography, 144
    use with sulfur-selective
        chemiluminescence detection, 3–8
    use with supercritical fluid
        chromatography, 64
Flame sulfur chemiluminescence detector:
    applications of, 37–38
    comparison with flameless type, 43–45
    disadvantages of, 42–43
Flavors, oxygenated hydrocarbon
    determination in, 115–116
Flow injection analysis, of nitrite ion in
    aqueous samples, 77–80
Fluorescent compounds, detection by
    peroxyoxalate chemiluminescence,
    228–229
Fluorine, gas chromatography/atomic
    emission detection of, 147, 149, 151,
    152, 154
Fluorine-induced chemiluminescence:
    detector design, 10
    efficiency of, 9
    mechanism of, 8–9
    response characteristics and applications
        of, 10–12
    use in sulfur-selective detection, 8–12
Food:
    cadmium determination in, 172
    nitrite determination in, 80
    nitrosamine determination in, 82
    N-nitroso compound determination in,
        74–75

organonitrogen compound determination
    in, 85
Fossil fuels, sulfur determination in, 147
Fragrances, oxygenated hydrocarbon
    determination in, 115
Fructose, redox chemiluminescence
    analysis of, 93
Fuel oils, oxygenates in, 151–152
Fuels, oxygenated hydrocarbon detection
    in, 100–101
Furfurol, oxygen-sensitive detection of, 115

Gallium, gas chromatography/atomic
    emission detection of, 149, 163, 164
Gas chromatography:
    use with atomic plasma emission, 143–169
    use with flameless sulfur
        chemiluminescence detection, 43,
        48, 59–63
    use with nitric oxide/ozone
        chemiluminescence detection, 75, 82
    use with plasma mass spectrometric
        detection, 190–197
        low-pressure gas chromatography,
        192–195
    use with pyrochemiluminescence
        detection, 85
    use with redox chemiluminescence
        detector, 91, 92–93
Gas chromatography/atomic emission
    detection of elements, 143–169
    analytical results, 148–150
    detector for, 144–148
    instrumentation for, 144–150
    nonmetallic elements, 150–160
    using atmospheric pressure plasmas,
        153–160
    using reduced pressure plasmas, 151–153
Gasohol, redox chemiluminescence analysis
    of, 93
Gasoline:
    methyl-*tert*-butyl ether determination in, 92
    MMT determination in, 167
    oxygenated hydrocarbon determination
        in, 114, 115, 116, 123
    sulfur detection in, 60
    sulfur limits in, 59
Geochemistry, gas chromatography/atomic
    emission detection use in, 156, 157

Germanium, gas chromatography/atomic emission detection of, 149

Glycerol trinitrate, detection in blood plasma, 82

Gold:
chromatography/plasma mass spectrometry detection of, 178, 181–182, 187
drugs containing, 173, 181–182
speciation of, 173

Gold catalyst, use in nitric oxide detection method, 75

Hall electrolytic conductivity detector. *See* Electrolytic conductivity (Hall) detector

Halogenated hydrocarbons, chromatography/plasma mass spectrometry detection of, 179, 191, 193, 195, 201

Halogens:
gas chromatography/atomic emission detection of, 144, 147, 149, 151, 153
gas chromatography/plasma mass spectrometry detection of, 194

n-Heptane, cracking capacity of, 108

1-Heptanol, oxygen determination in, 109

Hexadecane, as reagent for hydrogen atmosphere flame ionization detector, 130

n-Hexane, sulfur compound detection in, 49–51

1-Hexanol, oxygen determination in, 109

High-performance liquid chromatography:
use in nitric oxide/ozone chemiluminescence detection, 82
use with peroxyoxalate chemiluminescence, 237
use with plasma mass spectrometric detection, 176–190

Hot dog extract, nitrite determination in, 80

Hydrocarbons:
combustion products of, 48
lack of interference from, in sulfur chemiluminescence detection, 51
oxygenated, oxygen detector for, 115
sulfur detection in, 37–38, 54

Hydrogen:
gas chromatography/atomic emission detection of, 149, 151, 153, 160

redox chemiluminescence detection of, 92

Hydrogen atmosphere flame ionization detector (HAFID):
collar-biasing experiments on, 132–134
computer modeling experiments on, 132, 139–141
instrumentation for, 129, 130
organometallic compound detection by, 127
reagents for, 130–131
wall-biasing experiments on, 134–139

Hydrogen peroxide:
detection by peroxyoxalate chemiluminescence, 230–235
detection by redox chemiluminescence, 92

Hydrogen sulfide, reaction with ozone, 40

Hydroxylamine, reaction with nitric acid, 94

Indium:
gas chromatography/atomic emission determination of, 163
on-line preconcentration of, 188

Inductively coupled plasma (ICP):
use for atomic plasma emission chromatographic detection, 144
use for plasma mass spectrometry, 175

Industrial plant effluents, lead compound determination in, 160–161

Iodide, reaction with nitric acid, 94

Iodine:
chromatography/plasma mass spectrometry detection of, 178, 179, 191, 193, 195
gas chromatography/atomic emission detection of, 149, 160
speciation of, 173

Iodine-131, use in nuclear medicine, 173

Iodo amino acids, speciation of, 179

Ion-exchange chromatography, use with plasma mass spectrometric detection, 184–186

Ion-pairing chromatography, use with plasma mass spectrometric detection, 179–182

Iron:
gas chromatography/atomic emission determination of, 149, 165
gas chromatography/plasma mass spectrometry determination of, 196–197

Iron (*cont.*)
   speciation of, 173–174
Isobutanol, oxygen determination in, 110
Isopropanol, oxygen determination in, 110,
   112
Isosorbide, detection in blood plasma, 82

Kerogen, gas chromatography/atomic
   emission studies on, 156, 158, 160
Kerosene, oxygenated hydrocarbon
   determination in, 115, 118
Ketones, redox chemiluminescence
   detection of, 92
Kidney stones, oxalic acid monitoring for, 235
Kjeldahl procedure, pyroluminescence
   method compared to, 84, 85

Laser-initiated fluorescence, in
   peroxyoxalate chemiluminescence, 236
Lavender, oxygen-sensitive detector studies
   on, 118
Lead and lead compounds, 173
   chromatography/plasma mass
      spectrometry determination of, 178,
      186, 189, 190, 191, 199, 202
   gas chromatography/atomic emission
      determination of, 149, 160
   on-line preconcentration of, 188
   speciation of, 173, 190
Lead poisoning, zinc protoporphyrin
   detection in, 179
Lemon oil:
   oxygen-sensitive detector studies on, 118
   redox chemiluminescence analysis of, 93
Lettuce extract, nitrite determination in, 80
Limonene, oxygen-sensitive detector
   studies on, 118, 120
Liquid chromatography:
   use with flame and flameless sulfur
      chemiluminescence detection, 66, 67
   use with redox chemiluminescence
      detector, 91, 93–95
Liquid-phase sulfur analyzer (LPSA), use
   in aqueous-phase oxidation for sulfur
   monoxide formation, 22
Lovelock, James E., Pittcon Lecture of
   (1994), 241–249
Lubricating oil, aromatic hydrocarbon
   detection in, 115

MacSimion™ computer program. *See*
   SIMION modeling experiments.
Manganese, gas chromatography/atomic
   emission detection of, 149, 165
Mass spectral analysis, of hydrogen sulfide
   combustion products, 47, 48
Mass spectrometer, use in gas
   chromatography, 144
Mercaptans, detection of, 63
Mercury and mercury compounds
   chromatography/plasma mass
      spectrometry determination of, 178,
      189, 199, 202
   detection in food, 178
   gas chromatography/atomic emission
      determination of, 147, 149, 153, 160, 161
   on-line preconcentration of, 188
   speciation of, 174
Metallocenes, gas chromatography/atomic
   emission determination of, 164, 165
Metalloporphyrins:
   chromatography/plasma mass
      spectrometry detection of, 179
   in crude oils, 165
Metalloproteins, chromatography/plasma
   mass spectrometry detection of, 188
Metallothioneins, trace metal
   determinations on, 186
Metal selectivity, in flame ionization
   detector, 127
Methamphetamine, impurity detection in,
   185–186
Methane:
   cracking capacity of, 108
   formation in oxygen-sensitive detector,
      101, 105–106
Methanizer, poisoning of, 113
Methanol, oxygen determination in, 109,
   110, 112
Methyl acetate, oxygen determination in, 110
(Methylcyclopentadienyl)manganese
   tricarbonyl. See MMT
Methyl ethyl ketone, oxygen determination
   in, 110
Methyl mercaptan:
   chemiluminescence response to, 23
   combustion products of, 48
Methylmercury:
   chromatography/plasma mass
      spectrometry detection of, 178
   toxicity of, 174

Methyl tert-butyl ether, oxygen
    determination in, 109, 110, 112
Methyl-tert-butyl ether, determination in
    gasoline, 92
Micellar liquid chromatography, use with
    plasma mass spectrometric
    detection, 182–183
Microwave-induced plasma (MIP):
    as ion source for atomic plasma
        emission spectroscopy, 144,
        145–146, 151
    as ion source for plasma mass
        spectrometry, 175, 176
MMT, in gasoline, 167
Monomethylarsinic acid (MMA), toxicity
    of, 172
Myochrysin:
    chromatography/plasma mass
        spectrometry determination of, 178,
        181–182
    as gold-containing drug, 173

Naphtha, oxygenated hydrocarbon
    determination in, 115, 117
Nickel:
    gas chromatography/atomic plasma
        emission detection of, 149, 164, 165
    gas chromatography/plasma mass
        spectrometry detection of, 196–197
Nitrate ion detection:
    in aqueous samples, 75–80
    in atmospheric particles, 77
    in seawater, 76–77
Nitric acid, nitric oxide generation from, 93–94
Nitric oxide:
    chemiluminescence signal from, 26
    redox chemiluminescence detector for,
        88–91
Nitric oxide/ozone chemiluminescence
    detection methods, 71–97
    for aqueous solutions, 75–80
    history of, 71–72
    instrumentation for, 74–75, 76
    reactions in, 72–74
Nitrite ion:
    chemiluminescence signal from, 26
    detection of
        in aqueous samples, 75–80
        flow injection analysis, 77–80
        in seawater, 76–77

reaction with nitric acid, 94
Nitrogen:
    gas chromatography/atomic plasma
        emission detection of, 144, 149, 151,
        153, 154, 160
    gas chromatography/plasma
        spectrometric detection of, 192
    organic compounds containing. See
        Organonitrogen compounds
Nitrogen compounds, as interference in
    sulfur chemiluminescence detection,
    26, 51
Nitrogen dioxide:
    nonthermal conversion to nitric oxide
        for detection, 86–88
    photolysis to nitric oxide for detection,
        88–89
Nitrogen oxides (NO_x), thermal conversion
    to nitric oxide for detection, 86–88
Nitrogen–phosphorus detector (NPD):
    use in element-selective gas
        chromatography, 144
    use in thermionic conversion of sulfur
        compounds, 19–22
Nitrogen-selective detector, advantages of, 41
Nitrogen species (reactive, NO_y), catalytic
    conversion to nitric oxide for
    detection, 89–91
N-Nitrosamines, detection of
    by reductive cleavage, 82–83
    by thermal energy analysis, 81–82
N-Nitrosoatrazine, detection of, by
    reductive cleavage, 82–83
N-Nitroso compounds, detection of, 74–75,
    76, 81–84
    by thermal energy analysis, 81–82
N-Nitrosodiethanolamine, detection of, by
    reductive cleavage, 82–83

1-Octanol, oxygen determination in, 109
n-Octyl mercaptan, response factor relative
    to thiophene, 52
Oil shale retort waters and gases:
    gas chromatographic-atomic emission
        studies on, 152, 153
    organonitrogen compound detection in,
        84, 85
Olefins, chemiluminescence signals from, 26
Orange oil, oxygen-sensitive detector
    studies on, 118

Organic compounds, gas chromatography/
plasma mass spectrometry detection
of, 196–197
Organometallic compounds, detection by
flame ionization, 127–142
Organonitrogen compounds, detection of,
75, 84–85
Organophosphorus compounds, gas
chromatography/atomic plasma
emission spectroscopy studies on, 156
Organosulfur compounds, gas
chromatographic/atomic emission
studies on, 156
Oxalates, detection by peroxyoxalate
chemiluminescence, 231–232
Oxygen:
gas chromatography/atomic emission
detection of, 149, 151, 153, 154, 160
gas chromatography/plasma mass
spectrometry detection of, 192
Oxygen-sensitive detector (O-FID), 99–126
analytical parameters for, 107
applications of, 113–121
cracking reactor of, 103–105, 113
cracking temperature for, 111–113
dynamic range of, 107
gas supply for, 106
linearity of, 108
methanizer of, 101, 105–106, 113
principles of operation, 101–103
reproducibility of, 110–111
response factors of, 109–110
selectivity of, 109
sensitivity of, 108–109
use with FID alone, 118, 120, 121
Ozone-induced chemiluminescence
mechanism of, 13–24, 40
use in sulfur-selective detection, 12–30

Pacific Ocean, nitric oxide determination
in troposphere over, 85–86
Palladium, gas chromatography/atomic
emission detection of, 149, 164, 165
Penicillamine, aqueous phase conversion
detection of, 28
Pentaerythritol tetranitrate, detection in
blood plasma, 82
2,4-Pentanedione, use in chelating ligand, 163
1-Pentanol, oxygen determination in, 109

2-Pentanol, oxygen determination in, 110
Peptides, sulfur-containing, aqueous phase
conversion detection of, 25
Perfume, oxygen-sensitive detector studies
on, 121
Peroxyacetyl nitrate (PAN):
in atmosphere, 85
reduction to nitric oxide for detection, 88, 89
Peroxyoxalate chemiluminescence, 209–240
analytical applications of, 227–236
background emission studies on, 216
detection based on background-signal
enhancement, 235
fluorescent compound detection by,
228–229
fluorescent-tagged compound detection,
229–230
kinetic studies on, 213–216
optimization of, 221–227
photochemical initiation studies,
217–221, 236
quenched method, 235–236
reaction mechanism for, 210–221
Pesticides:
element determination in, 148
organonitrogen compound detection in, 85
sulfur-containing, aqueous phase
conversion detection of, 25, 28
Petroleum and petroleum products:
element determination in, 155
flameless sulfur chemiluminescence
detection use on, 35, 36–37, 63
organonitrogen compound detection in, 84
oxygen detector use for, 114
sulfur determination in, 147
Pharmaceuticals:
organonitrogen compound detection in, 85
sulfur-containing, aqueous phase
conversion detection of, 25
Phosphorus:
chromatography/plasma mass
spectrometry detection by, 179, 191,
193, 194, 195
gas chromatography/atomic emission
detection of, 144, 147, 149, 151
Plasma detection, supercritical fluid
chromatography use with, 64
Plasma mass spectrometry/chromatography:
argon chloride interference elimination, 189
chromatographic detection by, multielement

time-resolved analysis, 188
comparison studies, 185–190
direct injection nebulization for, 189
gradient elution in, 188
on-line preconcentration of, 188
plasma sources for, 175–176
Plasma mass spectrometry/chromatograpy,
171–207
Pleistocene sediment, gas chromatographic-
atomic emission studies on, 156
Pollution, atmospheric nitrate as indicator
of, 77
Polyamines, detection by peroxyoxalate
chemiluminescence, 231–232
Polychlorinated biphenyls (PCBs), gas
chromatography/atomic emission
studies on, 151, 154
Polycyclic aromatic hydrocarbons (PAHs),
detection by peroxyoxalate
chemiluminescence, 228–229
1-Propanol, oxygen determination in, 109
Proteins, sulfur-containing, aqueous phase
conversion detection of, 25, 28
PTFE membrane, use in reversed-phase
liquid chromatography, 24
Putrescine, detection in cancer patient
fluids, 231
Pyrochemiluminescence, organonitrogen
compound detection by, 84–85
Kjeldahl method compared to, 84–85

Radio frequency-induced plasma (RFP)
detector, as atomic plasma emission
chromatographic detector, 144, 147
Rare earths, determination of, 171
Redox chemiluminescence detector:
applications of, 91–95
use in generation of nitric oxide for
determination, 91–95
Reductive cleavage, N-nitroso compound
detection by, 83
applications, 83–84
Refinery gas, redox chemiluminescence
analysis of, 93
Reversed-phase liquid chromatography,
use with plasma mass spectrometric
detection, 177–179
Rheumatoid arthritis, gold-containing
drug therapy of, 173

Ribose, redox chemiluminescence analysis
of, 93
Ridaura. See Auranofin

Saccharides, redox chemiluminescence
analysis of, 93
Saliva, nitrite determination in, 80
Sausage, nitrosamine determination in, 82
Seawater:
nitrite and nitrate ion determination in,
76–77
organonitrogen compound detection in,
84
organotin determination in, 162–163
Selective detectors, for sulfur
chemiluminescence detection, 35
Selenium:
chromatography/plasma mass
spectrometry detection of, 180
gas chromatography/atomic emission
detection of, 149, 153
speciation of, 174, 180
Silicon, gas chromatography/atomic
emission detection of, 147, 149
SIMION modeling experiments, using
hydrogen atmosphere flame
ionization detector, 132, 139, 140
Size exclusion chromatography, use with
plasma mass spectrometric detection,
182–183
Soils, chromium in, 173
Solganal:
chromatography/plasma mass
spectrometry detection of, 182
as gold-containing drug, 173
Solute property detectors, use in gas
chromatography, 143
Spermidine, detection in cancer patient
fluids, 231
Spermine, detection in cancer patient
fluids, 231
Sucrose, redox chemiluminescence analysis
of, 93
Sulfates, detection by aqueous phase
conversion, 25
Sulfides, detection by aqueous phase
conversion, 25
Sulfites, detection by aqueous phase
conversion, 25

Sulfones, aqueous phase conversion
    detection of, 25
Sulfoxides, aqueous phase conversion
    detection of, 25
Sulfur:
    chromatography/plasma mass
        spectrometry determination of, 179,
        191, 193, 194, 195
    combustion chemistry of, 38–40
    gas chromatography/atomic emission
        determination of, 144, 147, 148, 149,
        151, 156, 160
    gas chromatography/plasma mass
        spectrometry determination of, 192
    gas chromatography/sulfur
        chemiluminescence determination
        of, 53–54
Sulfur chemiluminescence detector (SCD):
    advantages of, 12, 41
    flameless type, 19, 30
    flame-type, 17
    principle of operation of, 38–42
Sulfur compounds:
    in crude oil, 36
    ozone-induced chemiluminescence
        detection of
        by aqueous phase oxidation, 22–30
        by catalytic thermionic conversion,
            19–22, 30
        by direct reaction with ozone, 13–15, 30
        by flame conversion, 15–19, 30
        by flameless combustion, 19, 30
        method comparison, 30
        nitrogen-phosphorus detector for,
            19–22, 30
    widespread distribution of, 1
Sulfur dioxide:
    combustion products of, 48
    redox chemiluminescence detection of, 92
Sulfur monoxide:
    chemiluminescent reactions of, 12
    formation of, 40, 43–44
        in sulfur chemiluminescence detector, 40
    ozone-induced chemiluminescence detection
        of, 12–30
Sulfur-selective chemiluminescence
        detection in chromatography, 1–33
    flame photometric detector use, 3–8
    fluorine-induced chemiluminescence, 8–12
    ozone-induced chemiluminescence, 12–30
    uses, 3

Supercritical fluid chromatography:
    use with flameless sulfur
        chemiluminescence detection, 43, 58
    use with plasma mass spectrometric
        detection, 198–202
    use with redox chemiluminescence
        detector, 91, 93–95
Surfatron, as microwave-powered atmospheric
        pressure plasma cavity, 146

TBA, oxygen determination in, 110, 112
Tellurium:
    chromatographic detection by plasma
        mass spectrometry, 186
    speciation of, 174
Tetraalkyllead, toxicity of, 173
Tetradecane, as reagent for hydrogen
        atmorphere flame
        ionization detector, 130
Tetra-n-butyltin, as reagent for hydrogen
        atmosphere flame ionization
        detector, 130
Thermal conductivity detector, 143
Thermal energy analysis, N-nitroso
        compound detection by, 81–82
Thiirane-1-oxide, sulfur monoxide from
        decompoisition of, 23
Thimerosal, chromatography/plasma mass
        spectrometry detection of, 178, 179
o-Thiocresol, response factor relative to
        thiophene, 52
Thiocyanate, reaction with nitric acid, 94
Thiols, aqueous phase conversion detection
        of, 28
Thiophane, response factor relative to
        thiophene, 52
Thiophene(s):
    detection of
        by gas chromatography, 63
        by supercritical fluid chromatography, 65
    sulfur compound response factors relative
        to, 52
Thiosulfate, reaction with nitric acid, 94
Time-resolved analysis, in
        chromatography/plasma mass
        spectrometry, 188
Tin and tin compounds:
    chromatography/plasma mass
        spectrometry detection of, 178, 180,
        183, 187, 188, 190, 191, 195, 197,

199, 200, 202 gas chromatography/
atomic emission detection of, 149,
160, 162–163
speciation of, 174, 187, 190
Titanium, gas chromatography/atomic
emission detection of, 149
Toluene, cracking capacity of, 108
Trace elements:
determination of, 171–172, 184, 186
on-line concentration of, 188
speciation of, 172–174
Transition metals and compounds, gas
chromatographic/atomic emission
determination of, 160–167
Trialkyltin compounds, toxicity of, 174
Triphenylarsine, gas chromatographic/
atomic emission studies on, 156, 158
Tuna fish:
methylmercury determination in, 178, 179
tin detection in, 180

Urine:
arsenic detection in, 172

carbon-13 detection in, 159, 160
gold drug detection in, 182
nitrosamine detection in, 83
UV absorbance, nitrite determination by, 80

Vanadium, gas chromatography/atomic
emission detection of, 149, 164, 165,
166, 190
Volatile aromatic hydrocarbons (PAH),
determination in brown coal, 115

Wastewater, redox chemiluminescence
analysis of, 93
Water:
methylmercury detection in, 11162
nitric oxide detection in, 75–80
nitrogen compound detection in, 75–80

Zinc, chromatography/plasma mass
spectrometry detection of, 186, 187, 188